SOME NEW ASPECTS OF COLLOIDAL SYSTEMS IN FOODS

Edited by **Jafar M. Milani**

Some New Aspects of Colloidal Systems in Foods

http://dx.doi.org/10.5772/intechopen.75145
Edited by Jafar M. Milani

Contributors

Cristina Coman, Carlos Bravo-Díaz, Sonia Losada-Barreiro, Pascual García-Pérez, Eva Lozano-Milo, Pedro Pablo Gallego, Concha Tojo, Shahira Ezzat, Mohamed Salem, Camillo La Mesa, Gianfranco Risuleo, Jafar Mohammadzadeh Milani

Notice

Statements and opinions expressed in the chapters are these of the individual contributors and not necessarily those of the editors or publisher. No responsibility is accepted for the accuracy of information contained in the published chapters. The publisher assumes no responsibility for any damage or injury to persons or property arising out of the use of any materials, instructions, methods or ideas contained in the book.

First published in London, United Kingdom, 2019 by IntechOpen
IntechOpen is the global imprint of INTECHOPEN LIMITED, registered in England and Wales, registration number: 11086078, The Shard, 25th floor, 32 London Bridge Street
London, SE19SG – United Kingdom
Printed in Croatia

British Library Cataloguing-in-Publication Data
A catalogue record for this book is available from the British Library

Additional hard copies can be obtained from orders@intechopen.com

Some New Aspects of Colloidal Systems in Foods, Edited by Jafar M. Milani
p. cm.
Print ISBN 978-1-78985-781-8
Online ISBN 978-1-78985-782-5

Meet the editor

Jafar M. Milani is an Associate Professor in the Department of Food Science and Technology at Sari Agricultural Sciences and Natural Resources University (SANRU), Iran. He has been associated with teaching and research of food processes for the last 12 years. He has bachelor and master degrees in food science and a Ph.D. in food technology. He has guided more than 25 research scholars and has published more than 30 peer-reviewed research papers and 2 book chapters. He has worked on food product development, rheology, structure, and functional properties of hydrocolloids for food applications. Dr. Milani's teaching experiences include the principle of food preservation, food packaging and physical properties of foods.

Contents

Preface

Many food products typically exist in one of the forms of colloidal systems, and colloid science provides a strong background to modern food science. It is concerned with the study of the physical chemistry of dispersed systems. The practical benefits of the science of colloids for the food industry are controlling the sensory and textural aspects of existing food products as well as formulating different structures using innovative combinations of new ingredients and methods.

The goal of this book is a brief description of some recent developments in food colloids. The emphasis is placed on understanding the dynamics of colloids, the stability of colloids, approaches of production, applications, and regulatory aspects. This volume is therefore appropriate for university researchers and food manufacturers.

In the first chapter, classification and stability of colloidal systems are described. Then, the new structured functional ingredients for the design of a novel colloidal matrix are reviewed. Finally, the recent advances of colloidal systems in foods are discussed.

Adding of free radical scavenging antioxidants is one of the practical approaches controlling the oxidative stability in food emulsions. In the second chapter, applications of plant antioxidants in colloidal systems, ranging from their structural features to their benefits in human health as well as their antioxidant role in controlling the oxidative stresses mainly in lipid-based emulsions, are discussed.

Nanoemulsions are very attractive due to the advances of nanotechnology in recent years. Nanoemulsions have been applied in functional foods and pharmaceutical industries. Subsequently, nanoemulsion production with novel techniques and their stability is very important. Chapter 3 is an overview of nanoemulsion terminology and formulation, and various approaches for the production of nanoemulsions. In addition, the application of nanoemulsions in the food industry is discussed in detail.

In Chapter 4, nanostructured dispersions and their applications in various areas of food science are discussed. Nanostructured dispersions are nanosized materials that can be inherently present in food or they can be formed as the result of food processing technologies. Such materials can lead to the development of new and innovative food products and ingredients. Exciting fields of applications of nanostructured colloids in food science include functional food ingredients, food additives, food supplements, food packaging, and nanosensors.

In Chapter 5, the role of electrostatic and static forces in food dispersions and their stability is discussed. Emphasis is on the combination of various energy terms responsible for particle/particle attraction or repulsion. These forces are significant in dispersion stabilization and macroscopic phase separation.

Noticeably, the science of food colloids is a broad subject and various topics highlighted in this volume do not cover all of them. Finally, it is my pleasure to acknowledge the authors for the valuable information presented in this book.

Jafar M. Milani
Department of Food Science and Technology
Sari Agricultural Sciences and Natural Resources University (SANRU)
Sari, Iran

Chapter 1

Introductory Chapter: Some New Aspects of Colloidal Systems in Foods

Jafar M. Milani and Abdolkhalegh Golkar

Additional information is available at the end of the chapter

http://dx.doi.org/10.5772/intechopen.85298

1. Introduction

Food products usually showed the colloidal systems as emulsions, foams, gels, and dispersions. They are multicomponent systems containing different types of ingredients. Therefore, researches about food colloid are important; but first, we must be answered to below questions before starting. (1) How the physical properties of structure, stability, and rheology are influenced by the ingredient composition and formulation conditions? (2) How the interactions between various kinds of dispersed entities (e.g., particles, droplets, and bubbles) and macromolecules (proteins and polysaccharides) influence behavior in bulk fluid phases and at solid and liquid interfaces (air-in-water or oil-in-water)?

For these aims, many types of research are annually carried out for the understanding of how different classes of food ingredients control the physicochemical mechanisms determining overall stability and textural properties. Information on these model systems determined the reliable data about these colloidal systems, but not very reliant, because in the food products numerous components exist which influenced the quality and stability of final products. One of the main objectives of the colloid-based approach is the control of biopolymer interactions with the objective of fabricating well-defined nanoscale structures for controlled destabilization of colloidal systems [1, 2].

It should be noted that based on many papers presented at the 13th European Food Colloids Conference, almost all researches are focusing on four main areas: (i) structure and rheology of protein gels; (ii) properties of adsorbed protein layers; (iii) functionality from protein-polysaccharide interactions; and (iv) oral processing of food colloids. But nowadays, the behavior of dispersed systems within the human digestive systems has emerged as a major topic of research interest. Another outstanding influence on future food colloids research has a strong biomedical emphasis (the topic of controlled release and nutrient delivery) [3]. The development of colloid-based strategies to control delivery of nutrients during digestion in gastrointestinal is very important.

In this chapter, the progress in the field of ingredients, microstructure, and stability of food colloidal systems are discussed. Moreover, the application of new structured functional ingredients for the design of novel colloidal matrix (such as multilayer interfaces, multiple emulsion, gel-like emulsions, and so on) are reviewed. In this further, we will discuss: (i) classification and functions of colloidal systems in food; (ii) types of colloidal systems in food; and (iii) stability of colloidal systems.

A colloid system is a type of mixture in which one part is dispersed constantly throughout another. Colloid systems are usually formed when one part is dispersed through another, but does not combine to form a solution. Therefore, there are many types of colloidal systems that depend on the form of the two parts mixed together.

A colloidal system contains two separate phases: a dispersed phase (or internal phase) and a continuous phase (or dispersion medium). The part which is dispersed is known as the dispersed phase and is suspended in the continuous phase. Colloidal systems in food can be classified into different groups based on the states of matter constituting the two phases. Food colloids are sols, gels, emulsion, and foam. For example, egg white foam is a simple colloid system. Air bubbles (disperse phase) are trapped in the egg white (continuous phase) resulting in a foam. The detailed classification of colloidal systems in food is shown in **Table 1** [4].

Food colloids give structure, texture, and mouth-feel to many different food products; for example, jam, ice cream, mayonnaise, etc. Food colloid contains hydrocolloid that provides thickening, gelling, emulsifying, and stabilizing properties in food products [5].

Food hydrocolloids are high molecular weight hydrophilic biopolymers used as functional constituents in the food processing to modify microstructure, texture, flavor, and shelf-life. The term "hydrocolloid" comprises all the numerous polysaccharides that are obtained from plants, seaweeds, and microbial sources, as well as modified biopolymers made by the chemical or enzymatic treatment of starch or cellulose. One of the key functional roles of food hydrocolloids is in the preparation of emulsions and in the control of emulsion shelf-life. Most

System	Minor phase	Major phase	Products
Sol	Solid	Liquid	Raw custard, unset jelly
Gel	Liquid	Solid	Jelly and jam
Emulsion	Liquid	Liquid	Mayonnaise, milk
Solid emulsion	Liquid	Solid	Butter, margarine
Foam	Gas	Liquid	Whipped cream, whisked egg white
Solid foam	Gas	Solid	Meringue, bread, cake, ice cream

Table 1. Colloidal systems in food.

hydrocolloids can behave as stabilizers (stabilizing additives) of oil-in-water emulsions, but sole a few of them can act like emulsifiers (emulsifying agents). The second functionality needs considerable surface activity at the oil-water interface, and therefore the capability to favor the development and stabilization of fine droplets throughout and next emulsification [6, 7].

2. Stability of colloidal systems

The most part of colloids are stable, but the two phases may separate during a period of time due to an increase in temperature or through physical force. Furthermore, they may become unstable after freezing or heating, especially if they contain an emulsion of fat and water. The details of the instability of food colloids are reviewed in further part.

2.1. Sols and gels

A sol can be defined as a colloidal dispersion in which a solid is the dispersed phase and liquid is the continuous phase. Gravy stirred custard and other thick sauces are some of the examples. While jelly is formed, gelatin is scattered into a liquid and heated to make a sol. As the solution cooks, protein molecules unwind developing a network which traps water and creates a gel.

If corn flour is mixed with water and heated, the starch granules absorb water till they rapture, after that starch granule disperses in the water and the mixture becomes more viscous and makes a gel after cooling. Additional types of gel are formed with pectin and agar. Pectin, a form of carbohydrate found in fruit, is used in the production of jam to help it set. Agar is a polysaccharide extracted from seaweed which is capable of forming gels. If a gel is allowed to stand for a time, it starts to "weep." This loss of liquid is known as syneresis. The proper ratio of the ingredients is necessary to achieve the desired viscosity of the sols at a certain temperature.

Sols can be transformed into gels as a result of a reduction in temperature. In pectin gels, the pectin molecules are a major phase and the liquid is the scattered phase, whereas in pectin sol, the pectin molecules are a minor phase and the liquid is a major phase. Sols can be made as an initial step in the creation a gel. Jams and jellies produced using pectin are traditional cases that make a sol before the preferred structure.

2.2. Emulsions

Numerous natural and processed foods involve either relatively or entirely as emulsions or have been in an emulsified form at some time through their fabrication containing milk, cream, butter, margarine, fruit beverages, soups, batters, mayonnaise, cream liqueurs, sauces, desserts, salad cream, ice cream, and coffee whitener.

Emulsion products exhibit a wide variety of different physicochemical and organoleptic characteristics in appearance, aroma, texture, taste, and shelf-life. The processing of an emulsion-based food product with specific quality features is influenced by the selection of suitable raw materials (e.g., water, oil, emulsifiers, thickeners, minerals, acids, bases, vitamins, flavors, colorants, etc.) and processing situations (e.g., mixing, homogenization, pasteurization, sterilization, etc.).

An emulsion involves two immiscible phases (typically oil or water), with one of the liquids scattered as fine sphere-shaped droplets in the other. A system which contains oil droplets dispersed in an aqueous phase is termed an oil-in-water or O/W emulsion (e.g., mayonnaise, milk, cream, soups, and sauces). A system which involves water droplets scattered in an oil phase is termed as a water-in-oil or W/O emulsion (e.g., margarine, butter, and spreads) [8].

Multiple (or double) emulsions are multipart liquid dispersion systems well-known too as emulsions of emulsions, in which the droplets of one scattered liquid (water-in-oil or oil-in-water) are more scattered in another liquid (water or oil, correspondingly), making W/O/W or O/W/O. The innermost scattered droplets (hereafter called inner droplets or just droplets, while the droplets of the multiple emulsion will be named, for simplicity, the drops) in the multiple emulsion are disconnected from the external liquid phase by a film of another phase.

Although multiple emulsions are an emerging technology, only a few industrial products based on multiple emulsions exist in the marketplace. The main application of multiple emulsions is a protection system for the controlled release of active compounds. In the food industry, W/O/W emulsions are able to increase the solubility of specific active materials, solubilize oil-insoluble ingredients, and serve as protecting liquid reservoirs for molecules sensitive to outside environmental reactivity including oxidation, light, and enzymes, and act as entrapment reservoirs for covering off flavors and odors.

Applications in the cosmetics trade include aqueous preparations that provide a good "feel" and slow release of active materials or flavors, deposition of water-soluble agents onto the skin from wash-off systems. Most applications are related to the pharmaceutical industry, such as enhancing the chemotherapeutic effect of anticancer drugs, drug immobilization, treatment of drug overdoses, and protecting insulin from enzymatic degradation. However, the size of the droplets and the thermodynamic instability is a significant drawback of this technology. It seems that double-emulsion technology can now be applied in various areas, mainly in food, cosmetics, and pharmaceuticals [9, 10].

Emulsions are suggested as carriers of plant antioxidants in food systems that are discussed deeply in Chapter 2. In fact, plant antioxidants due to the natural sources and health-promoting product are very attractive in food science. So, information about the structure of plant antioxidants, degradation of them in food systems, physical and chemical stabilities of these systems are important for study in the future. In Chapter 3, the application of nanoemulsion in food science is discussed. Nanoemulsion is very attracting due to the advances of nanotechnology in the recent years. Nanoemulsion has been applied in functional foods and pharmaceutical industries. Therefore, nanoemulsion production with a novel technique and its stability is very important.

In general, emulsions are thermodynamically unstable and therefore tend to breakdown over time due to various physicochemical mechanisms. Therefore, stabilizers are used in emulsion formulations for improving their long-term stability, such as emulsifiers, texture modifiers, ripening inhibitors, and weighting agents. Emulsifiers (such as small molecule surfactants, phospholipids, proteins, polysaccharides, and other surface-active polymers) are typically amphiphilic molecules that have both hydrophilic and hydrophobic groups on the same molecule.

The most important polysaccharide emulsifiers in food applications are Arabic gum, modified starches, modified cellulose, pectin, and some galactomannans. The role of the emulsifier is to adsorb at the surface of the freshly formed fine droplets and so prevent them from coalescing with the near droplets to form larger droplets again. Mayonnaise is an example of a stable emulsion of oil and vinegar, when egg yolk (lecithin) may be used as an emulsifying agent.

Emulsions are usually formulated by a single type of emulsifier. But, in some cases, the quality and functional properties of emulsions can be improved by using a combination of several emulsifiers rather than the individual alone. Each of them has a unique molecular and physicochemical characteristic that can be applied for modulation the interfacial properties of emulsion droplets.

Novel or improved functional attributes can often be obtained by using emulsifier mixtures rather than single emulsifiers, for example, enhancements in antioxidant activity, flavor encapsulation, nutraceutical delivery, or textural attributes. Due to the increasing demand for clean-label products, utilization of natural emulsifier's mixtures can be recommended [11]. In addition, the new technique to conjugate proteins with polysaccharide by Maillard reaction arising in the controlled dry heating between the ϵ-amino groups of proteins and the reducing end carbonyl groups of polysaccharides are established. The most remarkable characteristic of the resultant protein-polysaccharide conjugates is the outstanding emulsifying attributes which are preferred in comparison to commercial emulsifiers [12].

Moreover, wet-heating has been adopted to prepare protein-polysaccharide conjugate. Wet-heating mostly shortens the reaction time to only several hours at high temperature and short reaction time limits the Maillard reaction to initial stage to provide better browning control [13, 14].

2.3. Foams

Foams consist of small bubbles of gas (frequently air) scattered in a liquid, for example, egg white foam. As liquid egg white is whipped, air bubbles are included. The mechanical action leads albumen proteins to unfold and make a network, entrapping the air. If egg white is heated, protein coagulates and moisture is driven off. This creates solid foam, for example, a meringue. Ice cream, bread, and cake are other instances of solid foams.

3. Recent advances in food colloidal systems and recommendations

Recent interest from researches is the application of structural design principles for the fabrication of edible colloids with novel functional properties. This research activity is driven forward by an increasing recognition within the food industry of the value of colloidal systems as delivery vehicles for nutrients and flavor compounds.

In Chapter 5, the role of electrostatic and steric forces in food colloids and their stability are discussed. Based on biopolymer interaction, a combination of protein and polysaccharide functionality for production of the novel biopolymer with enhanced functional properties is deeply studied. Proteins and polysaccharides are two groups of hydrocolloids that are widely used in food formulations simultaneously. These macromolecules are known to play important physicochemical roles, such as thickening, stabilizing, gelling, emulsifying properties, etc., in food products. Interactions between two hydrocolloids play an important role in the structure and stability of processed foods and depend not only on the physicochemical properties of proteins or polysaccharides alone [15–17].

Nowadays, the intelligent manipulation of protein and polysaccharide interactions provides opportunities for the design of new ingredients and interfacial structures with applications in the food and pharmaceutical industries.

So, food scientists can control the microstructure, texture, and shelf-life of edible colloidal systems with attention to theirs. Protein-polysaccharide interactions could play a key role in the nanoscale engineering of novel foods designed to address the widespread health concerns associated with obesity problem and the release of specific nutrients [18].

In addition, nowadays multilayer interfaces are very interested in food industries. Multilayer interfaces in food colloids typically consist of adsorbed layers of proteins and polysaccharides made by the sequential or simultaneous deposition of oppositely charged macromolecules at the surface of emulsion droplets [19].

With the advancement of nanotechnology in different fields such as food industry, some researchers studied various nanoencapsulation techniques for controlled and protection of some bioactive ingredients including pharmaceuticals and food bioactive components with the high bioavailability. Based on the main applied ingredients/equipment for the formulation of encapsulation systems, nanocarriers are classified into five groups: (1) lipid-based nanocarriers (such as nanoemulsions, nanoliposomes, and nanolipid carriers); (2) nanostructured colloids nanocarriers (such as caseins, cyclodextrins, and amylose); (3) nanocarriers produced by special equipment such as electro-spinning/spraying, nanospray dryer, and micro/nanofluidics systems; (4) biopolymers nanoparticles nanocarriers (such as single biopolymer nanoparticles, biopolymer-biopolymer complexation, nanogels of alginates, whey, soy proteins, and chitosan, nanotubes, or nanofibrils); and (5) miscellaneous nanocarriers (such as nanoparticles made from chemical polymers, nanostructured surfactants, inorganic nanoparticles, and nanocrystals). Hence, it is possible to choose appropriate nanodelivery systems based on the solubility and predicted functionality of bioactive components.

In last few years, there has been many published studies on the nanoencapsulation of different food ingredients such as phenolic compounds and antioxidants, natural food colorants, antimicrobial agents and essential oils, minerals, flavors, essential fatty acids and fish oil, and vitamins [20]. Active compounds such as antioxidants and antimicrobials are added into the food formulation for aims of quality loss and microbial safety management. But there are limitations such as pro-oxidation in lipid foods and compliance of regulatory maximum allowable concentration. Therefore, controlled release packaging (CRP) is a novel technology that is applied for the package with release active compounds in a controlled trend to improve safety and quality for food products during storage. Research in controlled release packaging focused on released systems such as active compounds from the package, non-releasing antimicrobials or antioxidants, oxygen absorbers, and free-radical scavengers those grafted on to packaging materials [21].

One developing area in the application of colloidal dispersions is the manufacture of functional foods. Functional foods are becoming progressively favorite among consumers as the result of improved knowledge of functional components and their influence on human wellbeing and biological functions. The customers would like to overcome health problems such as cardiovascular problems and obesity through consuming foods rather than drugs. The plan of functional foods for the delivery of nutraceuticals and micronutrients is a great technological challenge. Colloidal delivery systems are actually found in nature. Casein, for example, is a very illustrative instance of a natural colloidal delivery system for calcium. In milk, calcium is cleverly "engineered" into porous casein colloidal elements of sizes lesser than approximately 500 nm [22]. In Chapter 4, the nanostructured colloids in various areas of food science are discussed.

Nanostructured colloids can be naturally present in food or they can be synthetically manufactured. Some examples of natural nanostructured colloids include casein micelles and β-lactoglobulin in milk, and in the case of synthetically manufactured colloids are metal oxide nanoparticles and clay. Synthetically manufactured nanostructures are added to enhance solubility, improve bioavailability, biologically active compounds protection, increasing shelf-life, color, flavor, and add nutritional value.

The industrial sciences have been of great attention to the development of new bio-based structures with potential in innovative applications. Structures with gel-like behavior are usually used in the cosmetic, pharmaceutical, and food industries for the aim of controlling the physical properties of final products. In the food industry, words like oleogels and organogels have been increasingly used. Oleogels are new emulsion-based structure that can be used to control phase separation and decrease the mobility and migration of the oil phase, providing solid-like properties without using high levels of saturated fatty acids as well as to be a carrier of bioactive compounds. In this area, it can be used as the food grade and bio-based structurants for producing edible oleogels with fat replacement and structure-tailoring functionality [23].

Author details

Jafar M. Milani* and Abdolkhalegh Golkar

*Address all correspondence to: jmilany@yahoo.com

Department of Food Science and Technology, Sari Agricultural Sciences and Natural Resources University (SANRU), Sari, Iran

References

[1] Dickinson E. Stabilising emulsion-based colloidal structures with mixed food ingredients. Journal of the Science of Food and Agriculture. 2012;**93**:710-721. DOI: 10.1002/jsfa.6013

[2] Milani JM, Golkar A. Health aspect of novel hydrocolloids. In: Razavi SMA, editor. Emerging Natural Hydrocolloids: Rheology and Function. Oxford, UK: Wiley; 2019. pp. 601-622. DOI: 10.1002/9781119418511.ch24

[3] Dickinson E. Food colloids research: Historical perspective and outlook. Advances in Colloid and Interface Science. 2010;**165**:7-13. DOI: 10.1016/j.cis.2010.05.007

[4] Manisha S. Colloidal Systems in Food: Functions, Types, and Stability. Available from http://www.biotechnologynotes.com/food-biotechnology/food-chemistry/colloidal-systems-in-food-functions-types-and-stability-food-chemistry/14096 [February 15, 2019]

[5] Milani JM, Maleki G. Hydrocolloids in food industry. In: Valdez B, editor. Food Industrial Processes: Methods and Equipment. Rijeka, Croatia: InTech; 2012. pp. 17-38. DOI: 10.5772/32358

[6] Dickinson E. Hydrocolloids at interfaces and the influence on the properties of dispersed systems. Food Hydrocolloids. 2003;**17**:25-39. DOI: 10.1016/S0268-005X(01)00120-5

[7] Dickinson E. Hydrocolloids as emulsifiers and emulsion stabilizers. Food Hydrocolloids. 2009;**23**:1473-1482. DOI: 10.1016/j.foodhyd.2008.08.005

[8] McClements DJ. Food Emulsions: Principles, Practice, and Techniques. 2nd ed. New York, USA: CRC Press; 2005. 609p

[9] Garti N, Benichou A. Double emulsions for controlled-release applications—Progress and trends. In: Sjoblom J, editor. Encyclopedic Handbook of Emulsion Technology. New York, US: Marcel Dekker; 2001. pp. 377-407

[10] Aserin A. Multiple Emulsions: Technology and Application. New Jersey, USA: Wiley; 2008. DOI: 10.1002/9780470209264. 326p

[11] McClements DJ, Jafari SM. Improving emulsion formation, stability, and performance using mixed emulsifiers: A review. Advances in Colloid and Interface Science. 2017;**251**:55-79. DOI: 10.1016/j.cis.2017.12.001

[12] Kato A. Industrial applications of Maillard-type protein-polysaccharide conjugates. Food Science and Technology Research. 2002;**8**:193-199. DOI: 10.3136/fstr.8.193

[13] Zhang Xi QJR, Li KK, Yin SW, Wang JM, Zhu JH, Yang XQ. Characterization of soy β-conglycinin–dextran conjugate prepared by Maillard reaction in a crowded liquid system. Food Research International. 2012;**49**:648-654. DOI: 10.1016/j.foodres.2012.09.001

[14] Golkar A, Nasirpour A, Keramat J. Improving the emulsifying properties of β-lactoglobulin–wild almond gum (*Amygdalus scoparia* Spach) exudate complexes by heat. Journal of the Science of Food and Agriculture. 2016;**97**:341-349. DOI: 10.1002/jsfa.7741

[15] Golkar A, Nasirpour A, Keramat J. β-lactoglobulin-Angum Gum (*Amygdalus scoparia Spach*) complexes: Preparation and emulsion stabilization. Journal of Dispersion Science and Technology. 2015;**36**:685-694. DOI: 10.1080/01932691.2014.919587

[16] Golkar A, Nasirpour A, Keramat J. Emulsifying properties of Angum Gum (*Amygdalus scoparia Spach*) conjugated to β-lactoglobulin through Maillard-type reaction. International Journal of Food Properties. 2015;**18**:2042-2055. DOI: 10.1080/10942912.2014.962040

[17] Goh KT, Sarkar A, Singh H. Milk protein-polysaccharide interactions. In: Thompson A, Boland M, Singh H, editors. Milk Proteins: From Expression to Food. 2nd ed. USA: Academic Press; 2014. pp. 387-419. DOI: 10.1016/B978-0-12-405171-3.00013-1

[18] Dickinson E. Interfacial structure and stability of food emulsions as affected by protein-polysaccharide interactions. Soft Matter. 2008;**4**:932-942. DOI: 10.1039/B718319D

[19] Dickinson E. Colloids in food: Ingredients, structure, and stability. Annual Review of Food Science and Technology. 2015;**6**:2.1-2.23. DOI: 10.1146/annurev-food-022814-015651

[20] Assadpour E, Jafari SM. A systematic review on the nanoencapsulation of food bioactive ingredients and nutraceuticals by various nanocarriers. Critical Reviews in Food Science and Nutrition. 2018;**8**:1-47. DOI: 10.1080/10408398.2018.1484687

[21] Chen X, Chen M, Xu C, Yam KL. A critical review of controlled release packaging to improve food safety and quality. Critical Reviews in Food Science and Nutrition. 2018;**19**:1-14. DOI: 10.1080/10408398.2018.1453778

[22] Velikov KP, Pelan E. Colloidal delivery systems for micronutrients and nutraceuticals. Soft Matter. 2008;**4**:1964-1980. DOI: 10.1039/B804863K

[23] Martins AJ, Vicente AA, Cunha RL, Cerqueira MA. Edible oleogels: An opportunity for fat replacement in foods. Food & Function. 2018;**9**:758-773. DOI: 10.1039/C7FO01641G

Chapter 2

Plant Antioxidants in Food Emulsions

Pascual García-Pérez, Eva Lozano-Milo,
Pedro P. Gallego, Concha Tojo,
Sonia Losada-Barreiro and Carlos Bravo-Díaz

Additional information is available at the end of the chapter

http://dx.doi.org/10.5772/intechopen.79592

Abstract

Addition of free radical scavenging antioxidants (AOs) is one of practical strategies controlling the oxidative stability in food emulsions. Attention has been directed toward AOs derived from natural plant extracts with the capacity to improve health and well-being due to lack of consumers' trust toward synthetic antioxidant in food. Nevertheless, antioxidant efficiency varies widely from one compound to another and the most abundant AOs in our diet are not necessarily those that have the best availability profile at the reaction place with free radicals. In this book chapter, we will provide a state-of-the-art summary of the uses of plant AOs in colloidal systems, ranging from their main structural features to their benefits for the human health and their antioxidant role in controlling the oxidative stress and, particularly, the oxidation of lipid-based food emulsions.

Keywords: antioxidants, oxidative stress, lipid oxidation, emulsion stability

1. Introduction

Oxygen plays a controversial role in life: its presence is essential for aerobic organisms and, at the same time, it has been extensively reported as a toxic agent. Such toxicity derives from its capacity to form free radicals, considered highly energetic, unstable compounds with the ability of reacting easily with other molecules because they have unpaired electrons in the outermost orbitals. Whatever is the initiating mechanism, once the free radicals are formed, they can react with a biologically relevant molecule such as lipids, proteins, DNA, etc., leading to significant molecular blockage, degradative oxidation, and/or cell damage [1]. Polyunsaturated fatty acids, PUFAs (see **Figure 1**) are especially susceptible to chemical oxidation.

linolenic acid	eicosapentaenoic acid	docosahexaenoic acid
18:3 n-3, **LA**	20:5 n-3, EPA	22:6 n-3, **DHA**

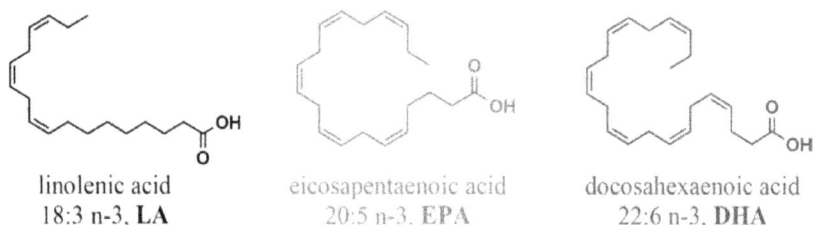

Figure 1. Chemical structures of some PUFAs.

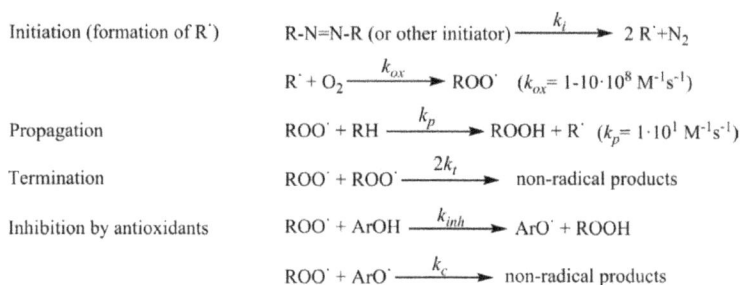

Initiation (formation of R˙) $\quad$ R–N=N–R (or other initiator) $\xrightarrow{\;k_i\;}$ 2 R˙+N$_2$

$\qquad\qquad\qquad\qquad\qquad$ R˙ + O$_2$ $\xrightarrow{\;k_{ox}\;}$ ROO˙ $\quad (k_{ox}= 1\cdot10\cdot10^8\ M^{-1}s^{-1})$

Propagation $\qquad\qquad\qquad$ ROO˙ + RH $\xrightarrow{\;k_p\;}$ ROOH + R˙ $\quad (k_p= 1\cdot10^1\ M^{-1}s^{-1})$

Termination $\qquad\qquad\qquad$ ROO˙ + ROO˙ $\xrightarrow{\;2k_t\;}$ non-radical products

Inhibition by antioxidants $\quad$ ROO˙ + ArOH $\xrightarrow{\;k_{inh}\;}$ ArO˙ + ROOH

$\qquad\qquad\qquad\qquad\qquad$ ROO˙ + ArO˙ $\xrightarrow{\;k_c\;}$ non-radical products

Figure 2. Main steps of the free radical oxidation of lipids and the inhibition by radical scavenger AOs. Average rate constant values for representative steps are also included. Constant abbreviations are found in the text. RH, lipid substrate; R˙, lipid radical; ROO˙, lipid peroxyl radical; ROOH, lipid hydroperoxide; ArOH, antioxidant; ArO˙ antioxidant radical.

In practical terms, lipid peroxidation (**Figure 2**) implies the reaction between a preexisting free radical and PUFAs during the initiation phase, whose reaction rate is given by the initiation constant, k_i, in which fatty acid radicals (R˙) are generated. Subsequently, these radicals are able to react rapidly with molecular oxygen (O$_2$) to produce fatty acid hydroperoxyl radicals (ROO˙), which constitutes the starting point for the subsequent propagation phase, driven by the rate of propagation constant, k_p, leading to oxidative stress. Finally, oxidative chain reactions may undergo a termination phase, characterized by its constant, k_t, where non-radical products are formed. Subsequently, ROO˙ radicals can be intercepted by AOs, throughout inhibition reactions, depending on their constant, k_{inh}. Furthermore, intercepted AOs may undergo termination reactions toward non-radical compounds production driven by the rate constant k_c, **Figure 2** [2].

Oxidative stress was first defined by Sies [3] as the lack of balance between the occurrence of reactive oxygen and nitrogen species (ROS and RNS, respectively) and the organism's capacity to counteract their action by the antioxidative protection systems. Since then, oxidative stress has been widely studied for decades, as it plays a key role on the etiology of several chronical diseases, i.e., diabetes, inflammation-related, neurodegenerative, and cardiovascular diseases, and cancer [1]. In order to overcome the deleterious effects attributed to oxidative stress, cells should maintain their redox homeostasis by enhancing *de novo* synthesis of AOs or by uptaking them from the diet or other exogenous sources.

2. Plant antioxidants

According to **Figure 2**, an efficient antioxidant is that molecule whose rate of trapping peroxyl radicals equals or overcomes the rate of formation of peroxyl radicals in the initiation step. Both rates depend on their intrinsic rate constants k and on the concentrations of reactants at the reaction site. Thus, the efficiency of a known amount of AO in a given reaction site depends on the value of the absolute rate constant for inhibition, k_{inh}, compared to the propagation rate constant, k_p, for reaction of the substrate with peroxyl radicals, e.g., the ratio k_p/k_{inh}. In general, AOs may protect against oxidation by a combination of various mechanisms. The predominant mechanism in a particular situation determines, to a great extent, the efficiency of the AO in inhibiting lipid oxidation.

Due to the high number of AOs reported to date and the plethora of reaction mechanisms that may be involved, the classification can be established according to multiple criteria. In a simple approach, AOs can be classified as follows: (1) according to their reactivity, (2) according to their origin, and (3) according to their structural features.

2.1. Functional classification of AOs

Based on functional features, AOs can be classified in two general groups: *primary* or *chain-breaking* AOs, responsible for the defense against ROS attack, by intercepting chain-propagating, O-centered, free radicals; and *secondary* or *preventive* AOs, that may prevent the attack of ROS on a substrate.

2.1.1. Primary or chain-breaking AOs

Primary (chain-breaking) AOs can trap two lipid radicals by donating a hydrogen atom to one radical and receiving an electron from another radical to form stable non-radical products. They inactivate free radicals by three main mechanisms, all playing important roles in determining radical scavenging activities depending on the particular environmental conditions: (1) transferring H-atoms to peroxyl radicals (hydrogen atom transfer mechanism, HAT), (2) electron transfer-proton transfer mechanism (SETPT), and (3) sequential proton loss-electron transfer mechanisms (SPLET) [4–6]. Probably the most common is the HAT mechanism, which involves the hemolytic cleavage of the O-H bond, converting them into harmless hydroperoxides and the oxidized antioxidant radical, which is less reactive with respect to $R^{\bullet}$, $RO^{\bullet}$, or $ROO^{\bullet}$ because of the delocation of the unpaired electron within their structures to form stable resonance hybrids.

2.1.2. Secondary or preventive AOs

Secondary AOs' function is closely related to lipid peroxidation, as it focuses on the interception of oxidative propagation processes after initiation. These agents may exert synergistic effects along with primary AOs as well, through several mechanisms [1]: (1) providing an acidic environment to stabilize primary AOs; (2) acting as hydrogen donors to regenerate primary AOs; (3) promoting metal chelation activity; and (4) quenching molecular oxygen with the aim of intercepting its reaction with oxidation-sensitive compounds.

Oxidized transition metal ions (mainly iron and copper) are one of the leading causes of the oxidative burst since their reduction implies the participation of O_2 and hydrogen peroxide (H_2O_2). Consequently, oxygen is enrolled in the initiation of redox reactions and could give rise to oxygen-, lipid-, or protein-derived radical formation. In this sense, chelators are able to interfere with metal oxidized ions to avoid the implication of O_2. Catechol or galloyl moieties-containing AOs have the ability of forming complexes with metals due to the presence of adjacent hydroxyl groups (—OH) within their structures, conversely to AOs bearing unique —OH groups (e.g., vanillic, syringic, and ferulic acids). A well-known chelating agent is citric acid, widely found in plant-derived foods, which is able to form stable coordination complexes with transition metals (with a typical stoichiometry of 1:1). Citric acid-metal ion complexes (Fe^{2+}, Cu^{2+}, Al^{3+}, etc.) formation is driven by their respective equilibrium constants, K_1, ranging from 10^3 to 10^5 M^{-1}. Specifically, citric acid Fe^{2+} complex binding constant has been reported to be $K_1 \approx 1600$ M^{-1} [2]. Moreover, AOs such as caffeic acid can also chelate metal ions and its reported binding constants, K_2, range from 10^0 to 10^2 M^{-1}. In this sense, the binding constant for the formation of caffeic acid—Fe^{2+} complexes—is $K_2 = 8$ M^{-1}, as it has been reported elsewhere [7, 8].

2.2. Origin-based classification of AOs

AOs can be also classified, according to their origin, into endogenous and exogenous, since total antioxidant capacity in biological systems involves endogenous antioxidant systems (mainly enzymes) and exogenous antioxidant compounds, proceeding from the diet.

Endogenous AOs comprise all the inner compounds and enzymatic systems acting like antioxidant agents under physiological conditions. As a general rule, antioxidant enzymes not only catalyze the synthesis or regeneration of previously oxidized AOs but also develop additional activities as free-radical scavengers and peroxide decomposers (see **Figure 3**). The three major enzymatic systems acting as AOs are superoxide dismutases (SOD), catalases (CAT), and glutathione peroxidases (GPX) [6]. Particularly, these enzymes show an important structural feature that is closely related to their antioxidant properties since they all contain transition metal ions within their active structures.

Besides antioxidant enzymes, inner antioxidant agents, such as GSH, coenzyme Q, and uric acid, may act as indirect scavengers of free radicals, metal ion sequesters, and oxidation-repairing agents [9]. Nevertheless, in most cases, such features are correlated to enzymatic systems.

Exogenous AOs (especially polyphenols) are mainly obtained from the diet. Polyphenols are considered one of the major compounds proceeding from plant secondary metabolism, since their distribution in the plant kingdom is wide. These compounds have been categorized as nutraceuticals due to their presence in edible plants and antioxidant-rich foods and drug supplements; therefore, they have gained much attention as preventive agents of several

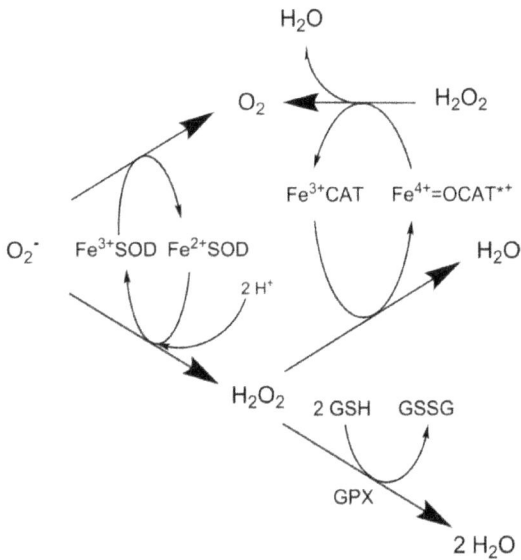

Figure 3. Overview of antioxidant enzymatic mechanisms: Fe-SOD: superoxide dismutase; Fe-CAT: catalase; GPX: glutathione peroxidase; GSH: glutathione; and GSSG: oxidized glutathione.

diseases worldwide. In this sense, neurodegenerative diseases, particularly Alzheimer's disease, and other oxidation-related diseases can be effectively prevented by polyphenol-rich foods intake. This way the regular consumption of polyphenols has been seen as an efficient antioxidant therapy against ROS due to the health benefits and anticancer effects attributed to such compounds [2].

2.3. Structural classification of plant-derived AOs

Plants have been used for centuries in traditional medicine for the effective treatment of several diseases. Such beneficial effects are a consequence of the biosynthesis of phytochemical compounds, derived from plant secondary metabolism in response to their environmental adaptation and protection. Thus, phytochemicals not only play a defensive role within plants but also many of them present additional features (acting as reproduction and environmental adaptation-related agents). Concerning dietary sources, the most common antioxidant phytochemicals found in vegetables and fruits are polyphenols and carotenoids [8]. Certain vitamins, which have also been isolated from several plant-based foods, show a strong antioxidant potential (see **Figure 4**).

2.3.1. Polyphenols

Polyphenols are structurally characterized by the presence of one or more aromatic rings bearing, and at least two hydroxyl groups in their chemical structures. As a consequence

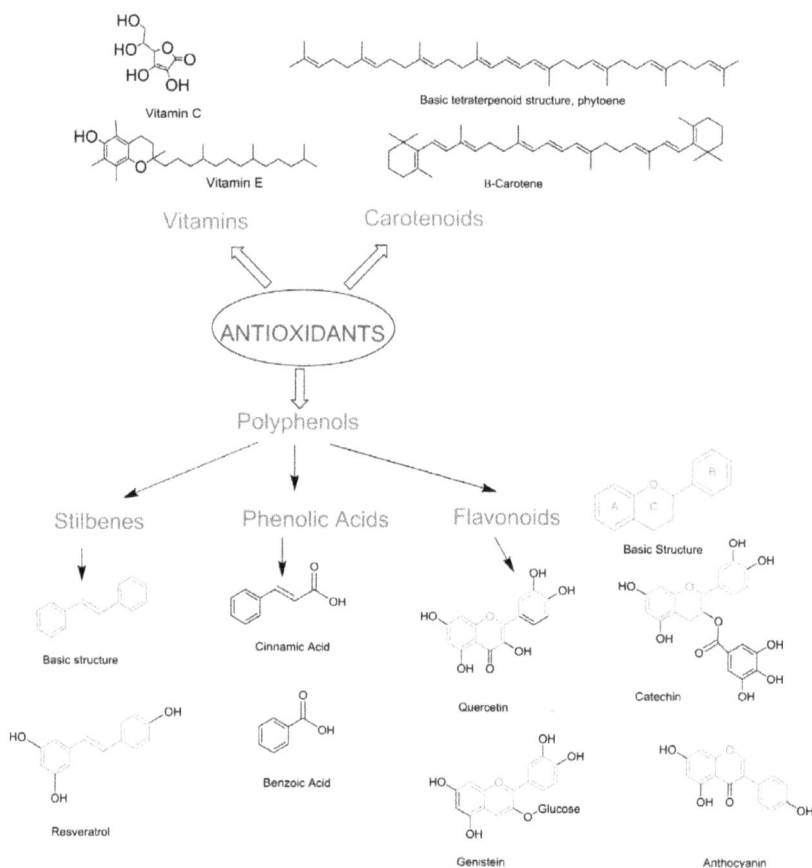

Figure 4. Typical classification of AOs and chemical structures of some representative AOs.

of their structural heterogeneity, polyphenols can be further subdivided into three groups, taking into account their presence in diet-derived products: flavonoids, phenolic acids, and stilbenes (see **Figure 4**).

2.3.1.1. Flavonoids

Flavonoids are plant secondary metabolites synthesized under oxidative conditions, due to their antioxidant properties as ROS scavengers, metal chelators, lipid peroxide decomposers, antioxidant enzymes inductors, and UV light absorbers [10]. Structurally, they are C6-C3-C6 benzo-γ-derivatives, containing phenolic and pyrane rings (**Figure 4**), and are broadly classified by the oxidation degree on their C-ring into: flavonols, flavones, flavanols (catechins), flavanones, isoflavones, and anthocyanins. However, only some of these groups achieve an effective antioxidant function throughout the diet.

2.3.1.2. Phenolic acids

Phenolic acids are plant secondary metabolites that contain at least one aromatic ring bearing one or more hydroxyl groups. They can be divided into two classes, according to the original acid they derive from: cinnamic acid (C6–C3) derivatives and benzoic acid (C6–C1) derivatives (**Figure 4**) [11]. The antioxidant activity attributed to these compounds is directly linked to the number of hydroxyl groups existing in their structure. Altogether, phenolic acids develop a pleiotropic antioxidant activity. The scavenging of oxygen and nitrogen-derived free radicals is the most significant feature which acts as chain-breaking AOs [12]. Just like flavonoids, phenolic acids are ubiquitous secondary metabolites found in various vegetables (artichoke and spinach), fruits (mainly citrus and berries), cereals, and coffee [11].

2.3.1.3. Stilbenes

Stilbenes are polyphenols containing a 1,2-diphenylethylene structural nucleus (**Figure 4**). Unlike other phenolic compounds, stilbene occurrence is limited within the plant kingdom and only discrete compounds have been studied in depth. In this sense, most of the studies concerning stilbenes have pointed at resveratrol (3,4′,5-trihydroxy-*trans*-stilbene, **Figure 4**) as the paramount compound belonging to this family. As an antioxidant agent, resveratrol possesses ROS and RNS scavenging activity and presents a strong influence upon the enhancement of antioxidant enzymatic systems [13]. Grapes, wine, and peanuts and their derived products are its principal sources [14].

2.3.2. Vitamins

For decades, many *in vitro* and *in vivo* studies highlighted the beneficial relation between the dietary consumption of vitamin-rich foods and the prevention of degenerative diseases, as a consequence of the antioxidant action of three vitamins: A, C, and E. Due to structural reasons, only vitamins C and E will be considered in this section, thus excluding vitamin A (a β-carotene-derived compound) to the next section.

Vitamin C (ascorbic acid, **Figure 4**) has been classically identified as a prominent antioxidant, due to its pleiotropic effects as free radical scavenger, metal chelator, and lipid oxidation inhibitor, with chain-breaking properties. An additional antioxidant feature of vitamin C is the ability of regenerating the oxidized form of vitamin E back to their active form, thus enhancing the synergistic effect between both compounds [15]. Fruits and vegetables, such as strawberries, papaya, kiwi, grapes, orange and their respective juices, pepper, tomato, and broccoli are considered their major sources [16].

Vitamin E covers a group of eight lipid-soluble molecules, derived from tocopherol and tocotrienol. Only α-tocopherol (**Figure 4**) has been found in significant amounts in dietary foods, such as edible oils and seeds [17]. The antioxidant efficiency of α-tocopherol is due to its chain-breaking properties, showing a strong specific affinity toward peroxyl radicals, exclusively [18].

2.3.3. Carotenoids

Carotenoids gained much interest in food chemistry as one of the major lipid-soluble groups of antioxidant compounds. Structurally, carotenoids are derived from phytoene, which is accepted as their tetraterpenoid precursor (**Figure 4**). However, only β-carotene (pro-vitamin A, **Figure 4**) and lycopene achieve a notable effect in dietary terms. They possess an enhanced effectiveness toward peroxyl radical scavenging and, additionally, their lipidic nature ensures an improved affinity to cell membranes, acting as protectors of these cell structures [15]. Due to their behavior as plant pigments, β-carotene is mainly found in highly pigmented fruits, such as apricots, carrots, and broccoli, while tomato is admitted as the major source of lycopene [19].

3. Food emulsions

Many natural and processed foods may be shown as examples of emulsion-type products. In the matter at hand, because of lipids that are highly hydrophobic and have a very low water solubility, they are usually incorporated into some kind of colloidal delivery system to make them dispersible in aqueous solutions. Emulsions are the main group of colloidal systems relevant for lipid-based foods [20]. Moreover, AOs are frequently added to lipid-based food emulsions because they are effective to hinder lipid oxidation. Over the past few years, there has been a growing emphasis on the understanding of the efficiency of AOs in emulsions, and it was found that antioxidant activity may vary largely depending on the composition of the emulsion [21]. Thereby, antioxidant activity is determined by a number of parameters such as interphase transport, surface accessibility, partitioning of AOs, and interaction of emulsifier with AOs. Advances in our understanding of the relationship between emulsion properties and antioxidant activity can be made through development of the principles and techniques of emulsion science, with the final purpose of improving the quality of the food production. The aim of this section is to give an introduction to the essential principles of emulsion science that are basic for understanding and manipulating the properties of food products.

Emulsion is a mixture of two immiscible liquids (usually oil and water), in which one liquid is dispersed as small spherical droplets in the other liquid. Foods such as milk, fruit juices, or mayonnaise, are composed by small droplets of oil dispersed in an aqueous solution (oil in water emulsion O/W). On the contrary, small droplets of water dispersed in a lipid medium are present in butter (water in oil emulsion W/O). They are all considered emulsions. The liquid present as discrete droplets is usually called disperse phase, and the another liquid is denoted by continuous phase.

The water phase in a food emulsion provides an unique environment, which not only supplies dissolving medium but also enhances interaction with many water-soluble components (protein, polysaccharide, vitamin, sugar, mineral, acid, base, preservative, etc.) [21, 22]. Also relevant is the oil phase since it contains aroma components at oil-water interface and dissolves several components, including AOs, oil-soluble vitamins, preservatives, and essential oils [20].

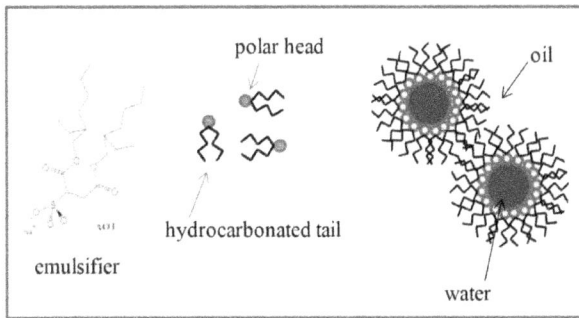

Figure 5. Representation of a location of the emulsifier at droplet surface.

3.1. Physical stability of emulsions

The molecules of the two immiscible liquids composing the emulsion are in direct contact with each other at the interface of each droplet. From a thermodynamical point of view, this arrangement is highly unfavorable. On the one hand, the entropy of the emulsion increases as the emulsion is formed due to increased entropy of mixing. However, this effect is not enough to compensate for the increased enthalpy, which is caused by the contact between hydrophilic and hydrophobic molecules. As a consequence, emulsions tend to separate the two liquids until the contact area between them is minimized (minimal free energy). Therefore, due to the presence of two immiscible phases, emulsions are thermodynamically unstable, that is, emulsions are vulnerable to break down over time by different process.

Although emulsions are thermodynamically unstable, it is possible to make it kinetically stable using an appropriate emulsifier. An emulsifier is usually a kind of molecule that consists of a water soluble hydrophilic part and an oil-soluble lipophilic part, as shown in **Figure 5**.

The addition of an emulsifier or surfactant to a mixture of water and oil stabilizes the emulsion because the emulsifier is arranged on the interface between the two phases [22]: the hydrophilic part of the emulsifier is anchored into water and its lipophilic part into oil. In this way, the emulsifier forms a film surrounding the surface of the droplets. It results in a reduction of interfacial tension, so emulsifiers or surfactants can be called surface-active compounds, which ensures kinetic stability in a certain period [23]. The choice of the emulsifier is crucial in the formation of the emulsion and its long-term stability (see **Figure 6**) [24].

3.1.1. Thermodynamic stability of emulsions

Droplet size and thermodynamic stability are one of the most characteristic features of an emulsion for classification as follows:

1. Macroemulsions: they constitute the most common emulsion type in foods, and they are thermodynamically unstable but kinetically stable. The droplet size ranges from 0.1 to 5 μm.

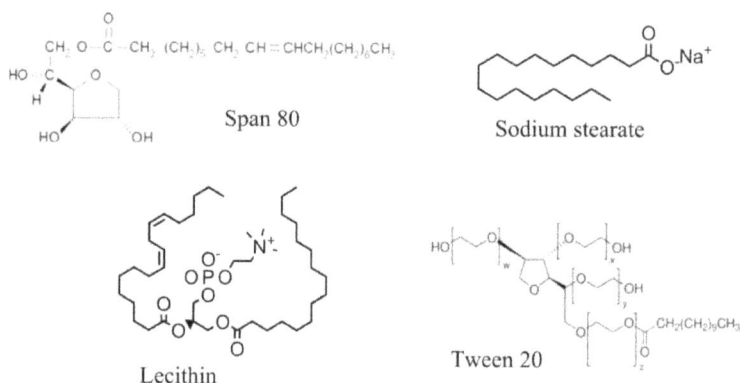

Figure 6. Common emulsifiers used to prepare food-grade emulsions.

2. Nanoemulsions: they are close to macroemulsions, but have a size range of 20–100 nm. They are only kinetically stable, so they are exposed to environmental degradation [25].

3. Microemulsions: droplet size is smaller than the previous ones (5–50 nm) and they are thermodynamically stable [26].

4. Multiple emulsions (W/O/W and O/W/O systems) can be described as emulsion of an emulsion. They are thermodynamically unstable dispersion systems [27].

Alternatively, emulsions may be classified according to HLB value (hydrophile-lipophile balance, see above) as hydrophilic (O/W type and HLB value of the water phase 9–18) or lipophilic (W/O type and HLB value of the oil phase 3–8).

3.1.2. Breakdown processes

One of the main goals of making food emulsions is to provide a stable and manageable source of food, whose properties do not significantly change until the product is consumed. Producing stable emulsions is already a challenge as it requires an in-depth understanding of interfacial physics, because emulsions are inherently unstable. To explain this, the fundamental nature of emulsions must be briefly considered.

Emulsions are usually prepared from the mixture of the two immiscible liquids by mechanical means, provided that the two liquids have no (or a very insufficient) mutual solubility. The average drop size in emulsions may grow with time until emulsions eventually separate into two liquid phases. The different breakdown processes are illustrated in **Figure 7**. These breakdown processes may occur on storage depending basically on the balance between attractive van der Waals forces and repulsive electrostatic (due to double layer) and steric forces.

The physical treatment involved in colloid stabilization is not simple, and it requires analysis of the various surface forces involved. DLVO[1] theory justifies why some colloids aggregate

---

[1]Derjaguin & Landau, Verwey & Overbeek.

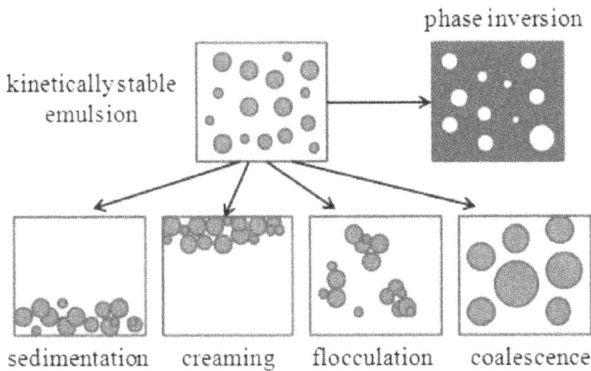

Figure 7. Scheme of breakdown processes that may occur in emulsions.

and others remain separately. Colloidal particles have electric charge and they are surrounded by ions with opposite charge, so an electric double layer is generated. The double layer causes electrostatic repulsion between droplets, which hinders their approximation. So double layer plays as a shield that provides kinetic stability. Colloidal particles come together if collision between two droplets is enough energetic to break the double layer and solvatation. If the particles strongly repel each other, the colloidal system will be stable.

A summary of the breakdown processes, details of each process, and methods of its prevention are given as follows.

Creaming and *sedimentation* take place when gravitational or centrifugal forces exceed the thermal motion of the droplets. If droplet density is lower than that of the medium, heavier droplets move faster to the top. On the contrary, they will move to the bottom when their density is larger than that of the medium. The closeness of the droplets favors breakdown of the interface. Eventually, the droplets can build up a close-packed arrangement at the top or bottom, giving rise to creaming or sedimentation, respectively. For a deeper discussion, see Refs. [28, 29]. The recovery of a creamed emulsion may be made by simply shaking or prevented by the following ways: (1) reducing the droplet size because the gravitational force is proportional to the cube of the droplet size; (2) increasing the viscosity of the continuous phase because it causes a slowdown of droplet movement; and (3) adding thickening agents (high-molecular-weight polymers such as carrageenans, alginates, etc., that hinder droplets motion and increase the viscosity in the continuous phase).

Coalescence is a growth process during which the emulsified droplets join together to form larger ones. Contrary to creaming, coalescence implies the irreversible fusion of droplets into larger ones, which implies disruption of the interdroplet liquid film. So the driving force for coalescence is the film fluctuations. The high mobility of molecules at the interface gives rise to fluctuations in the interface film, which can cause the film to break [30], and thus the two droplets spontaneously merge, causing coalescence. For a deeper study of coalescence, see Ref. [29].

Coalescence occurs in colloidal systems in which electrical repulsive effect is negligible, so it takes place particularly in O/W systems containing nonionic surfactants. Repulsive interactions can be modified by changing the charge at the surface, or by using a surfactant that provides a different thickness. Apart from that, coalescence is frequently caused by an incomplete covering of the droplets with surfactant molecules, so replace or increase surfactant can commonly avoid coalescence.

Flocculation is the process by which droplets (without changing droplet size) are aggregated into larger units. It occurs when there is not sufficient repulsion to keep the droplets apart to distances where the van der Waals attraction is weak. The van der Waals attraction is inversely proportional to the droplet-droplet distance of separation [29]. Flocculation is determined by the magnitude of attractive versus repulsive forces. On the contrary, coalescence is determined by the stability of the interdroplet film.

The van der Waals attractive forces depend on temperature, ionic strength, and charge of the interfacial layer [31], which strongly affects emulsion stability. Emulsions can be stabilized by electrostatic repulsion using ionic surfactants or by steric stabilization (adding large polymers to the surrounding aqueous phase).

Phase inversion is the process by which the dispersed phase and the medium are exchanged. As time goes by or due to a change in conditions, an O/W emulsion may turn to a W/O emulsion. Phase inversion usually takes place through a transition state including multiple emulsions. It can be minimized by choosing a suitable surfactant.

Finally, another process affecting emulsion stability is the *Ostwald ripening*, caused by the finite solubility of the liquid phases. Thus, liquids considered as being immiscible usually have no negligible solubilities. Specifically, in emulsions, curvature effects in the smaller droplets give rise to larger solubility than the larger ones. This difference in solubility between small and large droplets is the driving force for Ostwald ripening. The increase in solubility takes place when the droplet curvature increases, that is, when the droplet size decreases [32]. So the smaller droplets disappear and are deposited on the larger ones, resulting in larger droplets that grow at expenses of smaller ones.

There is a range of possibilities for modification in the properties of the emulsion which influence the stability and functional behavior of the colloidal system. Different protocols to ensure food storage stability can be found in Ref. [32]. All of them are based on the analysis of the following emulsion properties:

Droplet size distribution and *droplet concentration* are one of the most characteristic features of an emulsion not only due to the fact that most of the instability process are driven by droplet-droplet interactions, but also to the bulk properties (such as taste, color and texture), which depends mainly on these two parameters [22]. In particular, the speed of creaming depends on the effective particle size.

Composition of the stabilizing layer at the interface has an essential role to ensure emulsification and stability. The main component of the interfacial layer is the emulsifier or surfactant. The most used scale to classify emulsifiers is the hydrophilic-lipophilic balance (HLB), that is, a parameter relating molecular structure to interfacial packing and film curvature. The HLB value ranges from 0 to 20. An emulsifier with higher lipophilicity shows a lower HLB whereas higher hydrophilicity has high HLB, and the behaviors and functions to water depend on this HLB. This parameter was introduced by Griffin [33] in order to obtain an empirical Eq. (1) for nonionic alkyl polyglycol ethers (C_iE_j) based on the surfactant chemical composition, where E_j wt% and OH wt% are the weight percent of ethylene oxide and hydroxide groups, respectively.

$$HLB = (E_j \, wt\% + OH \, wt\%)/5 \qquad (1)$$

Also relevant can be the presence of solid particles and cosurfactant at the interface layer. Specifically, food emulsions frequently carry particulate material which is located at oil-water interface favoring emulsion stabilization [34]. Pickering-type food emulsions are emulsions consisting of droplets coated by a layer of adsorbed solid particles at the interface. The formation of O/W or W/O emulsions is determined by whether the particles are predominantly hydrophilic or hydrophobic [34].

The addition of a cosurfactant usually allows to enhance the effectiveness of surfactant. A cosurfactant is added to break up the assembling at the liquid/liquid interface so that it allows to attain lower interfacial tension. Furthermore, cosurfactants can be used to fine-tune the formulation, for example, by expanding the temperature or salinity range of microemulsion stability.

3.2. Chemical stability of the emulsions

As it was mentioned, an AO is efficient in inhibiting lipid oxidation of the emulsified foods if $r_{inh} \geq r_p$, where r_{inh} is the rate of trapping of the lipid radicals $R^\bullet$ and r_p is the rate for propagation step. However, many factors can affect the both rates. Both rates depend on their intrinsic rate constants k and on the concentrations of AOs at the reaction site. On the one hand, the radical scavenging activity of an AO depends on its chemical structure. On the other hand, the concentration of the AOs at the reaction site (that is, interfacial region in emulsions [35]) depends on the distribution of AOs between the oil, interfacial and aqueous regions of the emulsions, which in turn, depend on the physicochemical features of AOs and other parameters such as the nature of oil, the oil/water ratio, the electrical nature and the hydrophilic-lipophilic balance (HLB) of emulsifier, acidity and temperature [2]. Therefore, all these factors and their effects need to be taken into account to enhance the AO efficiency in depth.

3.2.1. Substituent effects

Substituents play a key role on the hydrogen atom donating capacity of AOs and understanding on their conformational, electronic, and geometrical characteristics is of vital significance to comprehend the relationship among AO structure and AO activity [36, 37]. The presence of electron donor groups, particularly at the ortho and/or para positions of the —OH group improve the scavenging activity of AOs due to lower the phenolic O—H bond dissociation enthalpy and higher reactivity with lipid radicals $R^\bullet$. For this reason, key structural features for a valuable radical scavenging activity are the position and number of hydroxyl groups attached to the aromatic ring and the presence of other functional groups such as alkyl chains containing C—C double bonds and C=O carbonyl groups, alkyl hydrocarbon chains among others. In general, it was found that rates for the reaction of catechols, 1,2-dihydroxybenzene and derivatives, with lipid radicals R• are higher than those for ortho-methoxyphenols due to the stabilization of the semiquinone radical formed from catechol. Catechols constitute the skeleton of many natural AOs such as flavonoid compounds (**Figure 4**). Structure of phenolics that allows conjugation and electronic delocalization, as well as resonance effects also can improve the radial scavenging activity of AOs.

3.2.2. Partitioning effects of AOs

Efficiency of the AOs depends not only on the AO nature but also on its concentration at the reaction site because AOs can be transferred between different regions (aqueous, oil and interface) of the food emulsions (see **Figure 8**), affecting their radical scavenging activity. Though the chemical properties and reactivities of relevant AOs toward free radicals are becoming well comprehended, it remains less clear how these properties translate into multiphasic systems. Thus, prediction of the efficiency of AOs in multiphasic systems such as food emulsions can become unclear since their partitions between the different regions were not explored [2, 38].

The physical impossibility of separating the interfacial region from the aqueous and oil regions of emulsions makes that any attempt to determine antioxidant distributions needs to be done in the intact emulsions, that is, without sample pretreatment. Application of the pseudophase kinetic model to emulsions provides a natural explanation, based on molecular properties, of the effects of a variety of parameters (nature and type of the oil, HLB, temperature, acidity, etc.) on the distribution of components between the oil, interfacial, and aqueous regions of emulsions prepared with edible oils [39, 40]. The reaction of choice was the reduction of a hydrophobic arenediazonium ion, whose reactive group is located in the interfacial region of the emulsions, and that can be monitored by a sampling method.

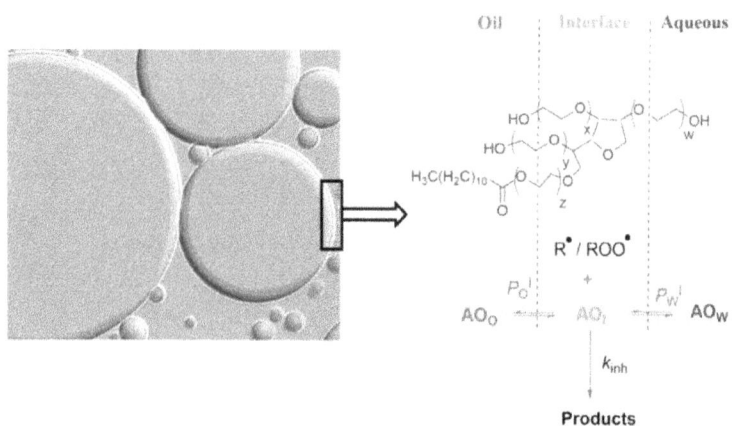

Figure 8. Left: optical microscope image of the droplets of an olive oil-in-water emulsion. Right: partition of AOs between the oil, interfacial (where lipid oxidation primarily occurs), and aqueous regions of the emulsion [38, 44].

Results obtained for a series of AOs (caffeic, gallic, protocatechuic acids, and hydroxytyrosol series) show that their distribution can be correlated with their antioxidant efficiency [35, 41–43]. This finding may have important consequences for the food industry because it opens the possibility of choosing the most efficient AOs for a particular food system on a scientific basis, resulting in an increase of the shelf-life of the product. Results should contribute to enhance current understanding of how antioxidant structure and physical location within the food system affect their efficiency and should provide basic information on the factors controlling antioxidant distributions and efficiencies, allowing a more rational selection of AOs and emulsifiers in food stabilization to maintain the organoleptic properties of foods.

3.2.3. Medium properties effects

The solvent properties of the reaction site affect both the intrinsic rate constant for the reaction between AO and free radicals, k_{inh}, and the partitioning of AOs between the different regions of the food emulsions. Among others:

- Emulsifier nature: the emulsifier electrical nature and HLB of the emulsifier can affect the concentration of AOs at the reaction site. Distribution results showed that the HLB of the emulsifier can modify the partition of moderate to high hydrophobicity AOs and the main parameter controlling the partition of AOs is the emulsifier concentration [45].

- Acidity of the aqueous region: the acidity of the medium can change substantially the partitioning of phenolic AOs. At the typical acidities of foods, phenolic AOs may be neutral or partially ionized and the ionic forms of the AOs are usually oil insoluble but much more aqueous soluble than the neutral forms, changing the partition of AOs and, as a consequence, the antioxidant efficiency [46].

- Oil nature: oxidation rates of monounsaturated fatty acids are much slower than those of polyunsaturated fatty acids. In this way, foods enriched with omega-3 can be seriously compromised by the oxidation of lipids due to their high degree of lipid unsaturation [35].

- Temperature: the temperature can affect lipid oxidation in many ways. It can produce not only variations in the rates for the reactions involved but also in the concentration of AOs in different regions of the emulsions due to changes in the solubility of the AOs [47].

- Oil/water (O/W) ratio: changing the O/W ratio of the emulsions significantly can affect the interfacial concentrations of very hydrophobic or hydrophilic AOs but not that of AOs of intermediate hydrophobicity [41].

4. Conclusions and future trends

Synthetic AOs have been employed for years, but because of the growing consumer interest in natural and health-promoting products, the industry is now challenged to offer new and efficient AOs derived from natural sources. Attention has, therefore, been directed toward the isolation of extracts of spices, herbs, and other plants rich in AOs because they have the capacity to minimize/inhibit oxidative degradation of biomolecules and thereby improve the quality and nutritional value of food. However, a careful choice of the AOs and the delivery system employed is crucial because it strongly affects their bioavailability and chemical reactivity against ROS. Their evaluation requires a wide range of experimental models from the simplest antioxidant assays in homogeneous solution to the biologically more relevant cellular systems. Many emulsion-based delivery systems for lipophilic compounds are regarded as one of the most promising techniques for transporting AOs to the target areas, deserving further investigations on the topics.

Acknowledgements

Financial support of the following institutions is also acknowledged: Xunta de Galicia (Programa REDES ED431D-2017/18), Ministerio de Educación y Ciencia (CTQ2006-13969-BQU) and University of Vigo. P. G.-P acknowledges the PhD Grant (FPU15/04849) to the Spanish Ministry of Education and S. L-B thanks the Postdoctoral Grant (POS-B/2016/012) to the Xunta de Galicia.

Author details

Pascual García-Pérez[1], Eva Lozano-Milo[1], Pedro P. Gallego[1], Concha Tojo[2], Sonia Losada-Barreiro[2]* and Carlos Bravo-Díaz[2]

*Address all correspondence to: sonia@uvigo.es

1 Plant Biology and Soil Science Department, Biology Faculty, University of Vigo, Vigo, Spain

2 Physical Chemistry Department, Chemistry Faculty, University of Vigo, Vigo, Spain

References

[1] Pisoschi AM, Pop A. The role of antioxidants in the chemistry of oxidative stress: A review. European Journal of Medicinal Chemistry. 2015;**97**:55-74

[2] Losada-Barreiro S, Bravo-Díaz C. Free radicals and polyphenols: The redox chemistry of neurodegenerative diseases. European Journal of Medicinal Chemistry. 2017;**133**:379-402

[3] Sies H, Cadenas E. Oxidative stress: Damage to intact cells and organs. Philosophical Transactions of the Royal Society of London. Series B, Biological Sciences. 1985;**311**:617-631

[4] Ross L, Barclay C, Vinqvist MR. Phenols as antioxidants. In: Rappoport Z, editor. The Chemistry of Phenols. West Sussex, England: J. Wiley & Sons; 2003

[5]   Shahidi F. Handbook of Antioxidants for Food Preservation. 1st ed. Oxford: Woodhead Pub.; 2015

[6]   Litwinienko G, Ingold KU. Solvent effects on the rates and mechanisms of reaction of phenols with free radicals. Accounts of Chemical Research. 2007;**40**(3):222-230

[7]   Andjelkovića M, Van Camp J, De Meulenaerb B, Depaemelaerea G, Socaciud C, Verloo M, et al. Iron-chelation properties of phenolic acids bearing catechol and galloyl groups. Food Chemistry. 2006;**98**:23-31

[8]   Oroian M, Escriche I. Antioxidants: Characterization, natural sources, extraction and analysis. Food Research International. 2015;**74**:10-36

[9]   Godic A, Poljšak B, Adamic M, Dahmane R. The role of antioxidants in skin cáncer prevention and treatment. Oxidative Medicine and Cellular Longevity. 2014;**2014**:860479-860485

[10]   Agati G, Azzarello E, Pollastri S, Tattini M. Flavonoids as antioxidants in plants: Location and functional significance. Plant Science. 2012;**196**:67-76

[11]   Heleno SA, Martins A, Queiroz MJRP, Ferreira ICFR. Bioactivity of phenolic acids: Metabolites versus parent compounds: A review. Food Chemistry. 2015;**173**:501-513

[12]   Kancheva VD. Phenolic antioxidants—radical-scavenging and chain-breaking activity: A comparative study. European Journal of Lipid Science and Technology. 2009;**111**: 1072-1089

[13]   Sirerol JA, Rodríguez ML, Mena S, Asensi MA, Estrela JM, Ortega AL. Role of natural stilbenes in the prevention of cancer. Oxidative Medicine and Cellular Longevity. 2016;**2016**:3128951-3128966

[14]   Wenzel E, Somoza V. Metabolism and bioavailability of trans-resveratrol. Molecular Nutrition & Food Research. 2005;**49**:472-481

[15]   Nimse SB, Palb D. Free radicals, natural antioxidants, and their reaction mechanisms. RSC Advances. 2015;**5**:27986-28006

[16]   Padayatty SJ, Katz-Arie WY, Eck P, Kwon O, Lee JH, Chen S, et al. Vitamin C as an antioxidant: Evaluation of its role in disease prevention. Journal of the American College of Nutrition. 2003;**22**(1):18

[17]   Galli F, Azzi A, Birringer M, Cook-Mills JM, Eggersdorfer M, Frank J, et al. Vitamin E: Emerging aspects and new directions. Free Radical Biology & Medicine. 2017;**102**:16-36

[18]   Niki E. Role of vitamin E as a lipid-soluble peroxyl radical scavenger: In vitro and in vivo evidence. Free Radical Biology & Medicine. 2014;**66**:3-12

[19]   Rao A, Rao L. Carotenoids and human health. Pharmacological Research. 2007;**55**:207-216

[20]   McClements DJ. Food Emulsions: Principles, Practices and Techniques. Boca Ratón: CRC Press; 2015

[21] Schwarz K, Huang SW, German JB, Tiersch B, Hartmann J, Frankel EN. Activities of anti-oxidants are affected by colloidal properties of oil-in-water and water-in-oil emulsions and bulk oils. Journal of Agricultural and Food Chemistry. 2000;**48**:4874-4882

[22] McClements DJ. Food Emulsions. Boca Raton, USA: CRC Press; 2005

[23] McClements DJ. Critical review of techniques and methodologies for characterization of emulsion stability. Critical Reviews in Food Science and Nutrition. 2007;**47**(7):611-649

[24] Tadros T. Applied Surfactants. Germany: Wiley-VCH Verlag; 2005

[25] McClements DJ, Rao J. Food-grade Nanoemulsions: Formulation, fabrication, proper-ties, performance, biological fate, and potential toxicity. Critical Reviews in Food Science and Nutrition. 2011;**51**:285-330

[26] Mehta SK, Kaur G. Microemulsions: Thermodynamic and Dynamic Properties. Rijeka, Croatia: InTech; 2011

[27] Dickinson E. Double emulsions stabilized by food biopolymers. Food Biophysics. 2011;**6**:1-11

[28] Sek J, Jozwiak B. Application of the continuity theory for the prediction of creaming phe-nomena in emulsions. Journal of Dispersion Science and Technology. 2015;**36**:991-999

[29] Tadros TF. Emulsion Science and Technology. Berlin: Wiley-VCH Verlag Chemie; 2013

[30] Dickinson E. An Introduction to Food Colloids. Oxford: Oxford University Press; 1992

[31] Israelachvili JN. Intermolecular and Surface Forces. London: Academic Press; 1992

[32] Weiss J. Emulsion Stability Determination. New York: John Wiley & Sons, Inc.; 2002

[33] Griffin WC. Classification of Surface-Active Agents by "HLB". The Journal of the Society of Cosmetic Chemists. 1949;**1**:311

[34] Dickinson E. Food emulsions and foams: Stabilization by particles. Current Opinion in Colloid & Interface Science. 2010;**15**:40-49

[35] Costa M, Losada-Barreiro S, Paiva-Martins F, Bravo-Díaz C. Physical evidence that the variations in the efficiency of homologous series of antioxidants in emulsions are due to differences in their partitioning. Journal of the Science of Food and Agriculture. 2017;**97**(2):564-571

[36] Vladimir-Knežević S, Blažeković B, Štefan MB, Babac M. Plant polyphenols as antioxi-dants influencing the human health. In: Rao V, editor. Phytochemicals as Nutraceuticals — Global Approaches to their Role in Nutrition and Health. London: InTech; 2012. DOI: 10.5772/2375

[37] Birben E, Sahiner UM, Sackesen C, Erzurum S, Kalayci O. Oxidative stress and antioxi-dant defense. World Allergy Organization Journal. 2012;**5**(1):9

[38] Decker EA, McClements DJ, Bourlieu-Lacanal C, Durand E, Figueroa-Espinoza MCL, et al. Hurdles in predicting antioxidant efficacy in oil-in-water emulsions. Trends in Food Science & Technology. 2017;**67**:183-194

[39] Bravo-Díaz C, Laurence RS, Losada-Barreiro S, Paiva-Martins F. Application of the pseudophase kinetic model to determining antioxidant distributions in emulsions: Why does dynamic equilibrium matters? European Journal of Lipid Science and Technology. 2017;**119**(12):12

[40] Bravo-Díaz C, Romsted LS, Liu C, Losada-Barreiro S, Pastoriza-Gallego MJ, Gao X, et al. To model chemical reactivity in heterogeneous emulsions, think homogeneous micro-emulsions. Langmuir. 2015;**31**:8961-8979

[41] Losada-Barreiro S, Bravo Díaz C, Paiva Martins F, Romsted LS. Maxima in antioxidant distributions and efficiencies with increasing hydrophobicity of gallic acid and its alkyl esters. The pseudophase model interpretation of the "cut-off effect". Journal of Agricultural and Food Chemistry. 2013;**61**:6533-6543

[42] Silva R, Losada-Barreiro S, Paiva-Martins F, Bravo-Díaz C. Partitioning and antioxida-tive effect of protocatechuates in soybean oil emulsions: Relevance of emulsifier concen-tration. European Journal of Lipid Science and Technology. 2017;**133**:379-402

[43] Almeida J, Losada-Barreiro S, Costa M, Paiva-Martins F, Bravo-Díaz C, Romsted LS. Interfacial concentrations of hydroxytyrosol and its lipophilic esters in intact olive oil-in-water emulsions: Effects of antioxidant hydrophobicity, surfactant concentra-tion, and the oil-to-water ratio on the oxidative stability of the emulsions. Journal of Agricultural and Food Chemistry. 2016;**64**:5274-5283

[44] Berton C, Ropers MH, Viau M, Genot C. Contribution of the interfacial layer to the protection of emulsified lipids against oxidation. Journal of Agricultural and Food Chemistry. 2011;**59**:5052-5061

[45] Losada-Barreiro S, Sánchez Paz V, Bravo-Díaz C. Effects of emulsifier hydrophile–lipo-phile balance and emulsifier concentration the distributions of gallic acid, propyl gal-late, and a-tocopherol in corn oil emulsions. Journal of Colloid and Interface Science. 2013;**389**:1-9

[46] Losada-Barreiro S, Bravo-Díaz C, Romsted LS. Distributions of phenolic acid anti-oxidants between the interfacial and aqueous regions of corn oil emulsions: Effects of pH and emulsifier concentration. European Journal of Lipid Science and Technology. 2015;**117**:1801-1813

[47] Losada-Barreiro S, Sánchez-Paz V, Bravo-Díaz C. Transfer of antioxidants at the inter-faces of model food emulsions: Distributions and thermodynamic parameters. Organic & Biomolecular Chemistry. 2015;**13**:876-885

Chapter 3

Nanoemulsions in Food Industry

Mohamed A. Salem and Shahira M. Ezzat

Additional information is available at the end of the chapter

http://dx.doi.org/10.5772/intechopen.79447

Abstract

A great attention has been received in the last few years for nanotechnology applications in food as well as pharmaceutical industries. People are looking for healthy and safe food worldwide. Therefore, researchers have been currently focusing on nanoemulsion technology that is particularly suited for the production of functional food. This chapter includes an overview about nanoemulsion terminology and formulation, various approaches for production of nanoemulsions which include high energy approaches such as high-pressure valve homogenization, microfluidizers and ultrasonic homogenization, and low energy methods such as spontaneous emulsification, phase inversion composition, phase inversion temperature and emulsion inversion point. In addition, the applications of nanoemulsions in food industry are discussed in detail.

Keywords: nanoemulsions, formulation, production approaches, food industry, applications

1. Introduction

Serious health-related problems contribute to the worldwide distribution of healthier, safer, and cost-effective food products. Additionally, functional foods were introduced as a tool to give an additional function to food. This can be achieved by increasing the production of existing biologically active molecules or adding new bioactive ingredients. Therefore, food products in addition to their nutritional value, they usually have health-promotion or disease prevention values. Nevertheless, it has become evident that the low bioavailability or inefficient long-term stability of these health-promoting products may not sustain their benefits. Subsequently, a great attention has been received in the last few years for nanotechnology in food applications. Nanoemulsions are one of the most interesting delivery systems in food industry. Nanoemulsion-based delivery systems improve the bioavailability of the encapsulated bioactive components and increase food stability [1].

Nanoemulsions are emulsions that have very small particle size [2]. They have some unique characteristics such as small size, increased surface area and less sensitivity to physical and chemical changes, making them ideal formulas in food industry [3, 4].

Food grade nanoemulsions are being increasingly used in for improving digestibility, efficient encapsulation, increasing bioavailability and targeted delivery [3–5]. The aforementioned advantages of nanoemulsions over the conventional emulsions increased the utility of nanoemulsions in food industry. The kinetic stability of nanoemulsions can be improved by incorporating stabilizers such as emulsifiers, ripening retarders, weighting agents or texture modifiers [3]. Emulsifiers such as small molecule surfactants (Tweens or Spans), amphiphilic polysaccharides (gum Arabic or modified starch), phospholipids (soy, egg or dairy lecithin) and amphiphilic proteins (caseinate or whey protein isolate) can be used in food industry to formulate nanoemulsions. Texture modifiers, substances that increase the viscosity such as proteins (whey protein isolate, gelatin or soy protein isolate), sugars (high-fructose corn syrup or sucrose), polysaccharides (carrageenan, xanthan, pectin, alginate) and polyols (sorbitol or glycerol) can be also used as stabilizers. Dense lipophilic materials such as brominated vegetable oil, sucrose acetate isobutyrate, ester gums can be used as a weighting agent to balance the densities of the liquids nanoemulsions [1, 3, 5–9].

In this chapter, we provide an overview on the terminology used in emulsions, formulation of nanoemulsions and diverse approaches for production of nanoemulsions. Additionally, we summarize the recent applications of nanoemulsions in food industry.

Emulsions are defined by International Union of Pure and Applied Chemistry (IUPAC) as "a fluid colloidal system in which liquid droplets and/or liquid crystals are dispersed in a liquid" [10]. If the continuous phase of the emulsion is an aqueous solution, the emulsion is oil-in-water and denoted by the symbol O/W, whereas, if the continuous phase is oil, the emulsion is referred to W/O (**Figure 1**) [10]. An emulsifier is a surfactant or surface-active agent, a substance that lowers the surface tension and/or the interfacial tension [10].

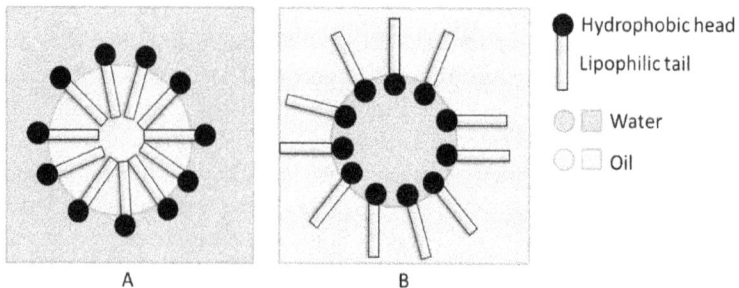

Figure 1. Schematic representation of oil in water (O/W, A) and water in oil (W/O, B) emulsions.

Nanoemulsions are emulsions that have a particle size at the nanometer range (20–500 nm) [2, 5, 6, 11]. Nanoemulsions have major differences in size, shape and stability from the classical macroemulsions and microemulsions [5]. While microemulsions are thermodynamically stable, both macroemulsions and nanoemulsions are thermodynamically unstable [5, 11].

2. Formulation of nanoemulsions

A typical nanoemulsion consists of a water phase, an oil phase and an emulsifier [5]. When present in small amounts, an emulsifier facilitates the formation of emulsions by decreasing the interfacial tension between the oil and water phases [5]. Additionally, emulsifiers aid the stabilization of nanoemulsions [11]. Formation and stabilization of nanoemulsions depend largely on the physico-chemical properties of the three aforementioned constituents.

O/W nanoemulsions have the greatest application in commercial products [9]. The particles in O/W nanoemulsion have a core-shell-type structure with a shell of surface-active amphiphilic material covers a core made of lipophilic material.

2.1. Oil phase

The oil phase used to prepare food-grade nanoemulsions can be formulated from a variety of nonpolar molecules, such as free fatty acids (FFA), monoacylglycerols (MAG), diacylglycerols (DAG), triacylglycerols (TAG), waxes, mineral oils or various lipophilic nutraceuticals [9]. TAG oils extracted from soybean, safflower, corn, flaxseed, sunflower, olive, algae or fish are the most commonly used in nanoemulsions primarily due to their low cost and nutritional value [9]. Physical and chemical characteristics of the oil phase such as viscosity, water solubility, density, polarity, refractive index and interfacial tension and chemical stability greatly influence the properties of nanoemulsions [1, 3, 5–8].

2.2. Aqueous phase

The aqueous phase used to prepare food-grade nanoemulsions can be formulated from water with a variety of polar molecules, carbohydrates, proteins, acids, minerals or alcoholic cosolvents [9]. The selection of the aqueous phase has a great impact on the physicochemical properties of the produced nanoemulsion.

2.3. Stabilizers

Stabilizers influence the long-term stability of nanoemulsions; therefore, the selection of the appropriate stabilizer is one of the most important factors to consider for the proper production of nanoemulsions. Various kinds of stabilizers are added to improve the long-term stability of nanoemulsions, and they are summarized in **Table 1** [1, 3, 5–9]. Stabilizers can be emulsifiers, ripening retarders, texture modifiers and weighting agents. Emulsifiers are

Stabilizers	Function	Examples
Emulsifiers	Single emulsifier or combination of emulsifiers are added to stabilize emulsions by increasing their kinetic stability	1. Small molecule surfactants (**Table 2**) 2. Amphiphilic polysaccharides (gum Arabic or modified starch) 3. Phospholipids (soy, egg or dairy lecithin) 4. Amphiphilic proteins (caseinate or whey protein isolate)
Ripening retarders	Hydrophobic substances that stabilize nanoemulsions by retarding or inhibiting Ostwald ripening	1. Mineral oil 2. Ester gum 3. Long-chain triglyceride
Texture modifiers	Substances that increase the viscosity of nanoemulsions	1. Proteins (whey protein isolate, gelatin or soy protein isolate) 2. Sugars (high-fructose corn syrup or sucrose) 3. Polysaccharides (carrageenan, xanthan, pectin, alginate) 4. Polyols (sorbitol or glycerol)
Weighting agents	Substances that balance the densities of the liquids nanoemulsions	1. Brominated vegetable oil 2. Sucrose acetate isobutyrate

Table 1. Kinds of stabilizers added to improve long-term stability of nanoemulsions.

Small molecule surfactants	Type	Examples	Remarks
Ionic surfactants	Negatively charged	Sodium lauryl sulfate (SLS) Diacetyl tartaric acid ester of mono- and diglycerides (DATEM) Citric acid esters of mono and diglycerides (CITREM)	• Used with low- and high-energy approaches. • They can cause irritation.
	Positively charged	Lauric arginate	
Nonionic surfactants	Sugar esters	Sucrose monopalmitate Sorbitan monooleate	• Used with low- and high-energy approaches. • They have low toxicity. • No irritability.
	Polyoxyethylene alkyl ethers (POE)	Brij-97	
	Ethoxylated sorbitan esters	Tweens 20 and 80 Spans 20, 40, 60 and 80	
Zwitterionic surfactants	Positively and negatively charged groups	Phospholipids (lecithin)	pH influences the net positive, negative or neutral charge

Table 2. Examples of small molecule surfactants added to nanoemulsions.

the most common stabilizers added in nanoemulsions. Emulsifiers of different kinds may be added such as phospholipids, small molecule surfactants, polysaccharides, and proteins. Examples of small molecule surfactants are listed in **Table 2**.

3. Approaches for production of nanoemulsions

Nanoemulsion is a nonequilibrium system which needs external or internal energy source to be successfully formed [12]. Nanoemulsions can be fabricated using many approaches that can be classified as high-energy or low-energy approaches.

The used techniques for the production of nanoemulsion has a great effect on the droplet size and consequently affect the stability mechanisms of the emulsion system through operating conditions and composition. Generally, preparation of nanoemulsions applies lower concentrations of surfactant (5–10%) than the microemulsion (20% and higher) [13].

Mechanical devices which can produce strong disruptive forces are used in high-energy approaches to mix and disrupt oil and water phases allowing the formation of tiny oil droplets [2, 14–16]. On the other hand, low energy approaches depend on the spontaneous formation of tiny oil droplets in the oil-water-emulsifier mixed systems when either the solution or the environmental conditions such as temperature and composition are changed [14, 17–21]. The used approach in nanoemulsion formation, together with the operating conditions, and the composition of the system affect the size of the formed droplets. In this section, we have a brief overview on the most commonly used high-energy and low-energy approaches for nanoemulsion formation.

3.1. High-energy emulsification

The preparation of nanoemulsions by high-energy methods is strongly dependent on the surfactants used as well as the functional compounds in addition to the quantity of energy applied [1]. Accordingly, the nanoemulsions formed by high-energy approaches constitute a natural method for the preservation of the nanoemulsions against formulation modification such as addition of monomer, surfactant, cosurfactants [17].

High-energy methods employ mechanical devices to produce disruptive forces that can mix and disrupt oil and water phases leading to the formation of the tiny oil droplets, such devices as high-pressure valve homogenizers, microfluidizers, and sonication methods [14, 16]. To keep the droplets in spherical shapes, intense energies are applied in order to generate disruptive forces that exceed the restoring forces, and these restoring forces could be calculated by the Laplace Pressure: $\Delta P = \gamma / 2r$, which increases by reducing droplet radius (r) and increasing interfacial tension (γ) [22]. Generally, the droplet size produced by high-energy approaches is controlled by a balance between two opposing processes that occur within the homogenizer, which are the droplet disruption and droplet coalescence [23]. Smaller droplets can be obtained by increasing the homogenization intensity or duration, increasing the concentration of the used emulsifier or by controlling the viscosity ratio [14, 22, 24]. The small-est droplet size that can be obtained using certain high-energy device is governed by the flow and force profiles of the homogenizer, the operating conditions such as the energy intensity and duration of the process, the environmental conditions meaning the applied temperature, the sample composition which includes the oil type, emulsifier type and concentrations, and the physicochemical properties of the phases which means the interfacial tension and viscos-ity [14, 25, 26]. In more clear words, the droplet size decreases as the intensity or duration of energy increases, the interfacial tension decreases, the emulsifier adsorption rate increases, and the disperse-to-continuous phase viscosity ratio falls within a certain range ($0.05 < \eta D/\eta C < 5$) [12, 13, 27]. Production of small droplets depends on the extent of the $\eta D/\eta C$ range and the nature of the disruptive forces produced by the particular homogenizer used, that is, simple shear versus extensional flow. Thus, the smaller the droplet radius, the more difficult is to break them up further.

High-energy approaches are the most suitable methods for the production of food-grade nanoemulsions as they can be applied to various types of oils such as triacylglycerol oils, flavor oils, and essentials oils as the oil phase as well as different emulsifier types such as proteins, polysaccharides, phospholipids, and surfactants. Even so, the size of the formed particles is strongly dependent on the oil characteristics and the used emulsifier. For instance, it is easier to produce very small droplets when flavor oils, essential oils or alkanes are used as the oil phase because they have low viscosity and/or interfacial tension [9]. Now we present the most commonly used devices in high-energy approaches.

3.1.1. High-pressure valve homogenization

In high-pressure valve homogenization, first a very high pressure is applied on the mixture and then it is pumped through a restrictive valve (**Figure 2**). The very fine emulsion droplets are generated by the very high shear stress [28, 29]. High-pressure and multiple passes are necessary to produce the required droplet size [9]. The combination of intensive disruptive forces such as shear stress, cavitation, and turbulent flow conditions can break the large drop-lets into smaller ones [30]. Production of conventional emulsions with small droplet sizes in food industry is commonly done using high-pressure valve homogenizers [22, 31]. Some of the food nanoemulsions prepared by high-pressure valve homogenization technique is β-carotene, thyme oil, and curcumin nanoemulsions [32–34].

These devices are more suitable for reducing the size of the droplets in preexisting coarse emulsions than in making emulsions directly from two separate liquids [9]. To describe the operation in the high-pressure valve homogenizer, the coarse emulsion is produced by the high shear mixer and is then passed into a chamber by the pump through the inlet of the high-pressure valve homogenizer and then forced through a narrow valve at the end of the chamber on its forward stroke. The coarse emulsion particles are broken down into smaller ones by a combination of intense disruptive forces when it passes through the valve. Different nozzle designs are available to increase the efficiency of droplet disruption [9].

The droplet size produced using a high-pressure valve homogenizer usually decreases as the number of passes and/or the homogenization pressure increases. It also depends on the

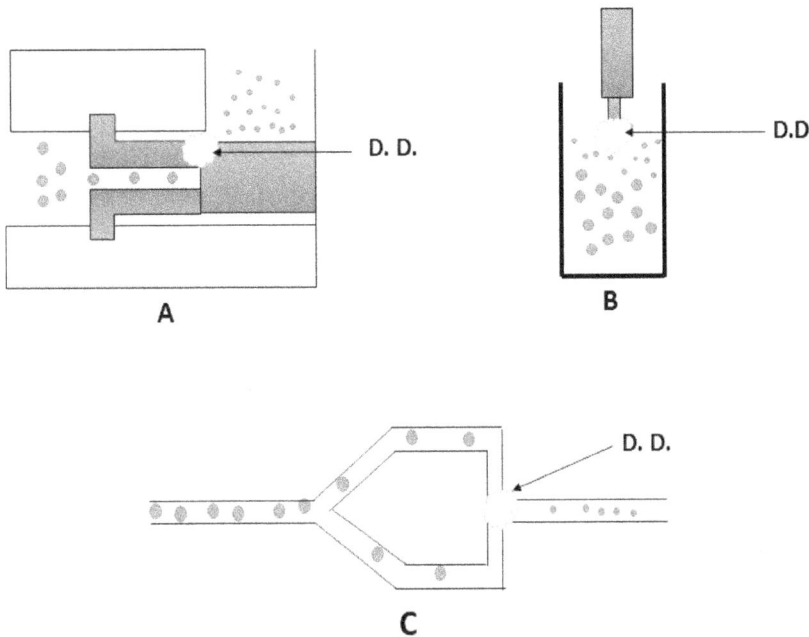

Figure 2. Schematic representation of three devices utilized in high-energy approach for production of food-grade nanoemulsions: A. high-pressure valve homogenization; B. ultrasonic homogenization; C. microfluidizer; and D.D. droplet disruption.

viscosity ratio of the two phases (usually oil and water) being homogenized. As mentioned before, small droplets can only usually be produced when the disperse-to-continuous phase viscosity ratio falls within a certain range ($0.05 < \eta D/\eta C < 5$) [12, 13, 27]. Moreover, sufficient emulsifier is required to cover the surfaces of the new droplet formed during homogenization, and the emulsifier should be rapidly adsorbed on the droplet surfaces to prevent recoalescence [23].

As a summary, to obtain the required droplet size in nanoemulsions, we need to operate at extremely high pressures and to use multiple passes through the homogenizer. Even then, high emulsifier levels, low interfacial tensions, and appropriate viscosity ratios are required to obtain droplets less than 100 nm in radius [9].

3.1.2. Microfluidizer

This device is similar in design to high-pressure valve homogenizer in that it employs high pressure to force the premix of emulsion through a narrow orifice to facilitate the disruption of droplet but differs only in the channels in which the emulsion flows (**Figure 2**). The emulsion flowing in a microfluidizer is divided through a channel into two streams, each passes through a separate fine channel, and then both streams are redirected into interaction chamber, in which they are exposed to intense disruptive forces leading to highly efficient droplet disruption [3]. Increasing the homogenization pressure, number of passes, and emulsifier concentration can efficiently reduce the droplet size formed. McClements and Rao have practically proved that the logarithm of the mean droplet diameter decreased linearly as the logarithm of the homogenization pressure increased for both ionic surfactant and a globular protein (β-lactoglobulin). But it could be noticed that this relation was appreciably steeper for the surfactant than for the protein, and this could be explained by the slow rate of the protein to adsorption to the droplet surfaces, with the formation of a viscoelastic coating which inhibits further droplet disruption [9].

In addition, there is an optimum range of disperse-to-continuous phase viscosity ratio, which facilitates the formation of small droplets [14]. But this relation is highly affected by the surfactant used, for the ionic surfactant mean droplet diameter decreases when the viscosity ratio decreases. On the other hand, the mean droplet size is hardly affected by viscosity ratio when a globular protein was used as an emulsifier, which may be also attributed to the relatively slow adsorption of the protein and its ability to form a coating that inhibits further droplet disruption [9].

Microfluidizers have been extensively used for the preparation of pharmaceutical products as nutraceutical emulsions, food and beverages such as homogenized milk in addition to the production of flavor emulsion [9]. Nanoemulsions of various bioactive compounds such as β-carotene and lemon grass essential oil were prepared using microfluidization technique [35–37].

3.1.3. Ultrasound homogenizer

When two immiscible liquids in the presence of a surfactant are subjected to high-frequency sound waves (frequency > 20 kHz) using sonicator probes that contain piezoelectric quartz crystals that expand and contract in response to an alternating electrical voltage, this causes strong shock waves produced in the surrounding liquid by the tip of the sonicator placed within the liquid (**Figure 2**). The mechanical vibrations lead to the formation of liquid jets at high speed, the collapse of the micro-bubbles formed by cavitation generates intense disruptive forces that lead to droplet disruption and the formation of emulsion droplets of nano size (70 nm). The emulsion should spend sufficient time within the region where droplet disruption occurs to ensure efficient and uniform homogenization [9, 16, 25, 38]. Practically, the droplet size decreases when the intensity of ultrasonic waves, sonication time, power level, and emulsifier concentration increase [25, 39, 40]. The type and amount of the emulsifier used, as well as the viscosity of the oil and aqueous phases affect the homogenization efficiency [16, 23, 25, 40]. All the above parameters should be first optimized to produce nanoemulsions of the right droplet size. It is noteworthy to mention that sonication methods may lead to protein denaturation, polysaccharide depolymerization, or lipid oxidation during homogenization. Thus, this technology has not yet been proved as efficient for industrial-scale applications [3].

3.1.4. High-speed devices

Rotor/stator devices (such as Ultra-Turrax) do not provide a good dispersion in terms of droplet sizes when compared to other high-energy techniques. The efficiency of such devices when calculated was 0.1%, where 99.9% is dissipated as heat during the homogenization process, so the energy provided mostly being dissipated, generating heat [12, 13, 17].

3.2. Low-energy emulsification

The low-energy methods are dependent on the internal chemical energy of the system [13, 41]. The nanoemulsions here are formed as a result of phase transitions that occur during the emulsification process due to the change in the environmental conditions such as temperature or composition [20], applying constant temperature and changing the composition or using constant composition and changing the temperature [42–44]. The composition of the emulsion such as surfactant-oil-water ratio, surfactant type and ionic strength in addition to the environmental conditions temperature, time history and stirring speeds greatly affect the droplet size [17, 44].

Low-energy approaches can produce smaller droplet sizes than high-energy approaches; however, low-energy approaches can be applied to limited types of oils and emulsifiers. For example, proteins or polysaccharides cannot be used as emulsifiers; alternatively, high concentrations of synthetic surfactants should be used to form nanoemulsions by low-energy approaches. This factor limits the use of such approaches in many food applications [4, 9]. The low-energy approaches are listed in the next section and represented in **Figure 3**.

3.2.1. Membrane emulsification method

This technique involves the formation of a dispersed phase (droplets) through a membrane into a continuous phase (**Figure 3**). It requires less surfactant and produces emulsions with a narrow size distribution range than the high-energy techniques. Unfortunately, the low flux of the dispersed phase through the membrane is a strong limitation during scale-up of this method [29].

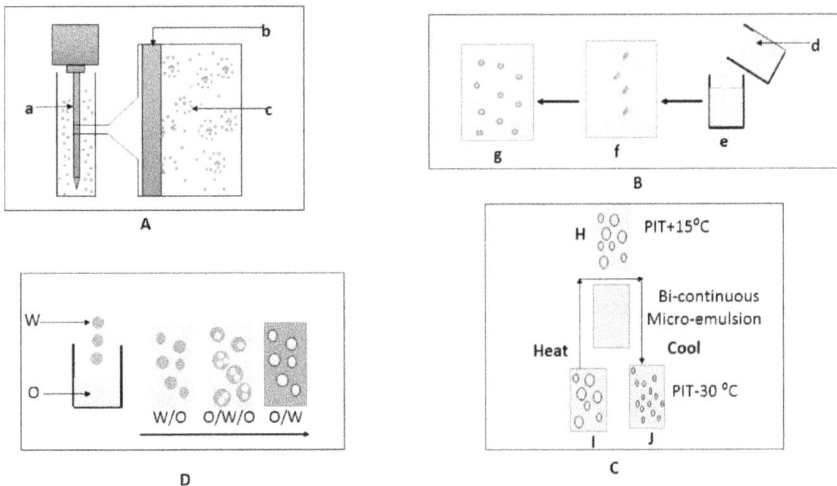

Figure 3. Schematic representation of four devices utilized in low-energy approach for production of food-grade nanoemulsions: A. membrane emulsification method; B. spontaneous emulsification method; C. phase inversion temperature method; D. emulsion inversion point method; a, rotating membrane; b, disperse phase; stabilized droplets of colloidal particles; d, surfactant and oil phase, e, aqueous phase; f, surfactant moves to water phase; g, oil in water emulsion; H. W/O emulsion; I. O/W emulsion; and J. O/W nanoemulsion.

3.2.2. Spontaneous emulsification method

This technique involves spontaneous formation of nanoemulsion as a result of the movement of a water miscible component from the organic phase into the aqueous phase when the two phases are mixed together (**Figure 3**) [17]. The organic phase is usually a homogeneous solution of oil, lipophilic surfactant and water-miscible solvent, and the aqueous phase consists of water and hydrophilic surfactant [19]. The spontaneous characteristic of this technique results from the initial nonequilibrium states when the two liquids are mixed without stirring. Accordingly, spontaneous emulsification is brought about by various methods such as diffusion of solutes between two phases, interfacial turbulence, surface tension gradient, dispersion mechanism or condensation mechanism. These mechanisms are highly influenced by the systems' compositions and their physicochemical features such as the physical properties of the oily phase and nature of the surfactants [19]. The size of the droplets produced can be controlled by varying the compositions of the two initial phases, as well as the mixing conditions [9].

There are many physicochemical mechanisms that can be utilized for spontaneous emulsification [45]. When two immiscible phases like water and oil are brought into contact with each other, and one of the phases contains a component that is partially miscible with both phases such as amphiphilic alcohol or surfactant. In this case, some of the components that are partially miscible with both phases will move from its original phase into the other one causing an increase in oil-water interfacial area, interfacial turbulence, and spontaneous formation of droplets. In this method, the variation in the compositions of the two initial phases, and the mixing conditions can control the size of the droplets produced.

McClements and Rao [9] compared the spontaneous emulsification method of producing nanoemulsions with the high-energy method named the microfluidizer. The surfactant-oil-water system used consisted of 15.4 wt% nonionic surfactant, 23.1 wt% medium-chain triglycerides (MCT), and 61.5% water, with the surfactant containing a 50:50 mixture of a hydrophilic (Tween 80) and lipophilic (Tween 85) surfactant. The microfluidization method produced droplets with a diameter of about 110 nm, whereas the spontaneous emulsification method could produce droplets with diameters around 140 nm. This simple experiment demonstrated that nanoemulsions could be produced using the spontaneous emulsification method, provided that the system composition was optimized, that is, surfactant, oil, and water contents.

This process itself increases entropy and thus decreases the Gibbs free energy of the system [17]. In pharmaceutical industry, the systems prepared by spontaneous emulsification method are referred to either as self-emulsifying drug-delivery systems (SEDDS) or as self-nano-emulsifying drug delivery systems (SNEDDS).

3.2.3. Solvent displacement

This method depends on the rapid diffusion of a water-miscible organic solvent that contains a lipophilic functional compound in the aqueous phase promoting the formation of nanoemulsions. This rapid diffusion enables the one-step preparation of nanoemulsion at low-energy input with high yield of encapsulation. At the end, the organic solvent is evaporated from the nanodispersion under vacuum [20, 21]. However, the use of this technique is limited to water-miscible solvents [21].

Another low-energy approaches are the phase inversion methods that use the chemical energy released as a result of phase transitions that occur during the emulsification. Nanoemulsions have been formed by inducing phase inversion in emulsion from a W/O to O/W form or vice versa by either changing the temperature in the phase-inversion temperature (PIT), the composition in phase-inversion composition (PIC) or emulsion-inversion point (EIP) [6].

3.2.4. Phase inversion temperature method

This method depends on that at a fixed composition and by changing temperature, the non-ionic surfactants changes their affinities to water and oil through the changes in the optimum curvature (molecular geometry) or relative solubility of nonionic surfactants [46, 47]. Using the PIT method, nanoemulsions are spontaneously formed by varying the temperature-time profile of certain mixtures of oil, water, and nonionic surfactant, thus nanoemulsions are formed by suddenly breaking-up the microemulsions maintained at the phase inversion point by a rapid cooling [48] or by a dilution in water or oil [17] the formed nanoemulsions are kinetically stable and can be considered as irreversible [3]. PIT also involves the controlled transformation of W/O emulsion to O/W emulsion or vice versa through an intermediate liquid crystalline or bicontinuous microemulsion phase [9].

The key for this phase inversion is the temperature-induced changes in the physicochemical properties of the surfactant (**Figure 3**). Here the molecular geometry of a surfactant is dependent on the packing parameter, $p = aT/aH$, where, aT is the cross-sectional area of the lipophilic tail-group and aH is the cross-sectional areas of the hydrophilic head-group [49].

In water, the surfactant molecules tend to associate with each other forming a monolayer due to the hydrophobic effects, and these monolayers have an optimum curvature that causes the most efficient packing of the molecules [49]. The packing parameter p determines the optimum curvature of the surfactant monolayer, when $p < 1$, the optimum curvature is convex and the surfactant favors the formation of O/W emulsions, for $p > 1$ the optimum curvature is concave favoring W/O emulsions, while for $p = 1$, monolayers have zero curvature, where surfactants do not favor either O/W or W/O systems and instead lead to the formation of form liquid crystalline or bicontinuous systems (**Figure 3**).

The relative solubility of surfactants in oil and water phases usually changes with temperature due to the physicochemical properties and packing parameter (p) of nonionic surfactants [50, 51]. At a particular temperature, the solubility of the surfactant in the oil and water phases is approximately equal, and this is known as phase inversion temperature or PIT at which an oil-water-surfactant system changes from an O/W emulsion to a W/O emulsion as the packing parameter equals unity ($p = 1$). At temperatures greater than the PIT ($\approx T > PIT +20°C$), the head group becomes progressively dehydrated and the solubility of the surfactant in water decreases, it becomes more soluble in oil, its $p > 1$, and the formation of a W/O emulsion is favored. When the temperature is decreased ($\approx T < PIT-30°C$), the head group of a nonionic surfactant becomes highly hydrated and tends to be more water soluble ($p < 1$), favoring the formation of O/W emulsions [9].

Above PIT, the surfactant molecules are being present predominantly within the oil droplets as they are more oil-soluble at this temperature. When this system is quench-cooled below the

PIT, the surfactant molecules rapidly move from the oil phase into the aqueous phase just like the movement of water-miscible solvent in the spontaneous-emulsification method, which leads to the spontaneous formation of small oil droplets because of the increase in interfacial area and interfacial turbulent flow generated. For this reason, Anton et al. [51] proposed that he formation of nanoemulsions by the PIT method has a similar physicochemical basis to the spontaneous emulsification method.

This process is characterized by being simple, prevents the encapsulated drug being degraded during processing, consumes low amounts of energy, and allows an easy industrial scale-up [17].

3.2.5. Phase inversion composition method

PIC method is very close to PIT method, but here the optimum curvature of the surfactant is altered by changing the formulation of the system, rather than the temperature [51]. For example, an O/W emulsion can be phase inverted to a W/O emulsion by adding salt as in this case the packing parameter increased and becomes greater than unity (p > 1) due to the ability of the salt ions to screen the electrical charge on the surfactant head groups [52]. Alternatively, a W/O emulsion containing a high salt concentration can be phase inverted to O/W emulsion by dilution with water in order to reduce the ionic strength below some critical level. Another PIC method for preparation of nanoemulsions is to change the electrical charge and stability of emulsions by changing the pH. The carboxyl groups of fatty acids are uncharged at low pH (pH < pKa) and have a relatively high oil solubility (p > 1), so they could stabilize W/O emulsions, but at high pH, they become ionized so they become more water-soluble (p < 1) and stabilize O/W emulsions. Consequently, nanoemulsions can be formed by increasing the pH of a fatty acid-oil-water mixture from below to above the pKa value of the carboxyl groups [41, 52].

3.2.6. Emulsion inversion point

This method involves changing the composition of the system at a constant temperature. In order to create kinetically stable nanoemulsions, the structures are formed through a progressive dilution with water or oil [17]. In EIP methods, the change from W/O to O/W or vice versa needs a catastrophic-phase inversion, rather than a transitional-phase inversion as with the PIC or PIT methods [53]. A transitional-phase inversion occurs when the characteristics of a surfactant are changed through adjusting one of the formulation variables, such as the temperature, pH, or ionic strength. A catastrophic-phase inversion occurs by changing the ratio of the oil-to-water phases while the surfactant properties remain constant. The emulsifiers used in catastrophic-phase inversion are usually limited to small molecule surfactants that can stabilize both W/O emulsions (at least over the short term) and O/W emulsions (over the long term) [9].

McClements and Rao [9] showed practically that increasing the amount of water in a W/O emulsion consisting of water droplets dispersed in oil with continuous stirring can cause the formation of additional water droplets within the oil phase at low amounts of added water;

however, once a critical water content is exceeded, the coalescence rate of water droplets exceeds the coalescence rate of oil droplets, and so phase inversion occurs from a W/O to an O/W system (**Figure 3**). Thus, the catastrophic-phase inversion is usually induced by either increasing (or decreasing) the volume fraction of the dispersed phase in an emulsion above (or below) some critical level.

The value of the critical concentration where phase inversion occurs, as well as the size of the oil droplets produced, depends on process variables, such as the stirring speed, the rate of water addition, and the emulsifier concentration [53].

4. Applications of nanoemulsions in food industry

Nanoemulsions have diverse applications such as drug delivery, pharmaceuticals, cosmetics, and food [5]. In this section, we focus on the applications of nanoemulsions in food industry. Nanoemulsions have been used as a suitable form to improve the digestibility of food, bioavailability of active components, pharmacological activities of certain compounds, and solubilization of drugs. Some applications are listed below.

4.1. Nanoemulsions and encapsulation of lipophilic components

One of the most important applications of nanoemulsions in food industries is the encapsulation of lipophilic components such as vitamins, flavors, and nutraceuticals [9]. Encapsulation is a useful tool to entrap a bioactive ingredient in a core or a fill within a carrier (coating, matrix, membrane, capsule, or shell) for improving the delivery of bioactive molecules within living cells [54].

This technology has many applications in food industry for masking the unpleasant taste or smell of some bioactive materials, increasing the bioavailability of some components, improving the stability of food ingredients, decreasing air-induced food degradation or decreasing the evaporation of food aroma [54]. One of the most interesting applications of encapsulation in food industry is probiotics. Probiotics are defined as microorganisms that provide health benefits when consumed in adequate amounts [55, 56]. Encapsulation of bioactive compounds in nanoemulsion-based delivery system was achieved for resveratrol (**Figure 4**) [57].

Nanoemulsions from food-grade ingredients are being increasingly utilized to encapsulate biologically active lipids such as Omega-3 fatty acids, polyunsaturated fatty acids (PUFAs) [9]. Omega-3 fatty acid supplementation may be protective effect against cancer, cardiac death, sudden death, cognitive aging, asthma, inflammation and myocardial infarction. α-Linolenic acid (ALA), an Omega–3 fatty acid, is one of two essential fatty acids together with linoleic acid (**Figure 4**). ALA is necessary for health and cannot be synthesized within the human body.

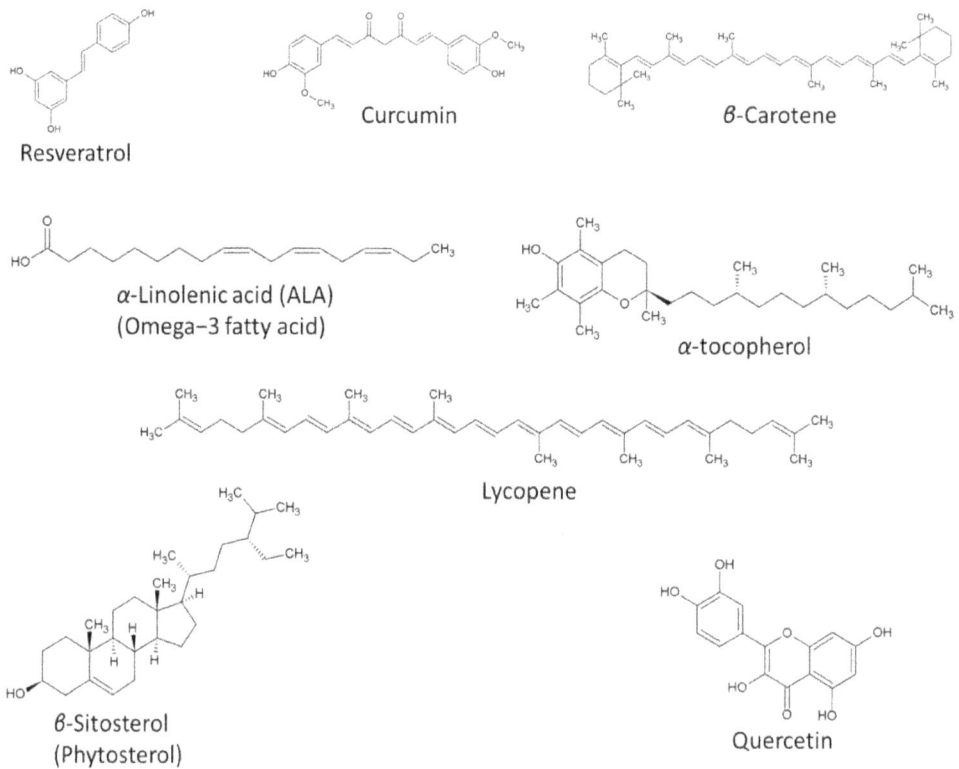

Figure 4. Applications of nanoemulsions-based delivery systems in food industry.

4.2. Nanoemulsions to improve drug bioavailability and pharmacological effects

Low bioavailability of some naturally occurring active compounds hinders their efficient pharmacological activities. Nanoemulsions have been used as a suitable form to increase bioavailability of natural extracts. Curcumin, 1,7-bis(4-hydroxy-3-methoxyphenyl)-1,6-heptadiene-3,5-dione (**Figure 4**), is a yellow-colored polyphenolic compound isolated from the rhizomes of turmeric (*Curcuma longa*, family Zingiberaceae) [58]. Curcumin has been used as a natural coloring agent health benefits such as anticarcinogenic, antioxidant, anti-inflammatory, and antimicrobial [59]. Curcumin nanoemulsions showed significant inhibition of 12-O-tetradecanoylphorbol-13-acetate (TPA)-induced inflammation [60]. However, low bioavailability hinders the efficiency of orally administrated curcumin. Flavored nanoemulsions have been prepared with improved curcumin digestibility compared to directly taken curcumin [60, 61].

Additionally, nanoemulsion formulation of oil-soluble vitamins such as alpha-tocopherol enhanced their oral bioavailability and pharmacological effects [62, 63]. α-Tocopherol, a type of vitamin E, is mainly present in olive and sunflower oils (**Figure 4**). Vitamin E supplements have important antioxidant, anticancer as well as cardiovascular protective activities.

Moreover, nanoemulsion preparations improved the bioavailability quercetin or methylquercetin [64]. Quercetin, a polyphenol from the flavonoid group of, has been found in many fruits, vegetables, leaves, and grains (**Figure 4**). Quercetin supplements have been promoted as antioxidant and anticancer.

4.3. Nanoemulsions to improve digestibility characters

Food digestibility is a measure of how much of food is absorbed by the gastro-intestinal tract into the bloodstream. Nanoemulsions have been used as a suitable form to improve digestibility characters of food and natural extracts.

β-Carotene is a red-orange pigment that is found in plants such as carrots and colorful vegetables. β-Carotene is a member of the carotenes, which are terpenoids (isoprenoids), biosynthesized from geranylgeranyl pyrophosphate (**Figure 4**). β-Carotene is the best-known provitamin A carotenoid. β-Carotene flavored nanoemulsion with improved digestibility has been applied [20, 33, 65].

4.4. Nanoemulsions to improve drug solubilization

Nanoemulsion formulation has been applied to increase the solubilization of phytosterols [66]. Phytosterols have been shown to lower the blood cholesterol, and therefore, they reduce the risk of coronary heart diseases. Among phytosterols, β-sitosterol has been isolated from many vegetables and fruits (**Figure 4**). Moreover, nanoemulsions formulas increased also the solubilization of lycopene [66]. Lycopene, a carotenoid pigment and phytochemical, has been found in tomatoes, other red fruits and vegetables (**Figure 4**). Lycopene has potential effects on prostate cancer and cardiovascular diseases.

5. Conclusion and future perspectives

Nanoemulsions have gained great attention and popularity during the last decade due to their exceptional properties such as high surface area, transparent appearance, robust stability, and tunable rheology. The most commonly known preparation approaches for nanoemulsions include high-energy approaches such as high-pressure valve homogenization, microfluidizers and ultrasonic homogenization, and low energy methods such as spontaneous emulsification, phase inversion composition, phase inversion temperature and emulsion inversion point. There is little understanding of the possible industrial relevance of many of these approaches as the physics of nanoemulsion formation is still semi-empirical and rational scale-up procedure have not been widely explored. The interest in nano-emulsion preparation and application is growing, but few of the numerous ideas reported become commercial final applications. Nanoemulsions are considered one of the most promising systems to improve solubility, bioavailability, and functionality of nonpolar active compounds. Food industry seeks to use these systems for the incorporation of the lipophilic functional compounds for the development of innovative food products. The application of nanoemulsions to food systems still poses challenges that need to be addressed both in terms of the production process, especially their cost, and of the characterization of both the resulting nanoemulsions and the food systems to which they will be applied in terms of product safety and acceptance.

Although nanoemulsions have potential advantages over conventional emulsions such as the preparation of transparent foods and beverages, their improved bioavailability, and physical stability. However, there are a number of regulatory aspects that should be overcome first to allow the wide applications of nanoemulsions.

First of all, most of the components used in formulation of nanoemulsions either in low-energy or high-energy approaches are unsuitable for widespread utilization within the food industry such as synthetic surfactants, synthetic polymers, synthetic oils, or organic solvents. Thus, food-grade ingredients such as flavor oils, triglyceride oils, proteins, and polysaccharides must be utilized in the formulation of food nanoemulsions as these ingredients are legally accepted, label-friendly and economically viable.

Second, in order to fabricate food-grade nanoemulsions on the industrial scale, suitable processing operations should be employed to obtain economic and robust products. Accordingly, many of the identified approaches which were developed in the research laboratories are not suitable for industrial production especially the low-intensity approaches, which could not be yet investigated in industrial scale production. At present, the high-intensity approaches only are utilized for production of nanoemulsions in the food industry.

Finally, there are certain safety concerns associated with the utilization of very small lipid droplets in foods. For example, the route of absorption, the bioavailability or potential toxicity of a lipophilic component encapsulated within nanometer-sized lipid droplets are considerably different from those dispersed within a bulk lipid phase. For these reasons, extensive studies are strongly needed in the area of nanoemulsion safety.

Author details

Mohamed A. Salem[1] and Shahira M. Ezzat[1,2]*

*Address all correspondence to: shahira.ezzat@pharma.cu.edu.eg

1 Department of Pharmacognosy, Faculty of Pharmacy, Cairo University, Cairo, Egypt

2 Department of Pharmacognosy, Faculty of Pharmacy, October University for Modern Sciences and Arts (MSA), Egypt

References

[1] Silva HD, Cerqueira MA, Vicente AA. Nanoemulsions for food applications: Development and characterization. Food and Bioprocess Technology. 2012;5(3):854-867

[2] Gutiérrez JM, González C, Maestro A, Solè I, Pey CM, Nolla J. Nano-emulsions: New applications and optimization of their preparation. Current Opinion in Colloid & Interface Science. 2008;13(4):245-251

[3] McClements DJ. Edible nanoemulsions: Fabrication, properties, and functional performance. Soft Matter. 2011;**7**(6):2297-2316

[4] Karthik P, Ezhilarasi PN, Anandharamakrishnan C. Challenges associated in stability of food grade nanoemulsions. Critical Reviews in Food Science and Nutrition. 2017;**57**(7):1435-1450

[5] Gupta A, Eral HB, Hatton TA, Doyle PS. Nanoemulsions: Formation, properties and applications. Soft Matter. 2016;**12**(11):2826-2841

[6] Solans C, Solé I. Nano-emulsions: Formation by low-energy methods. Current Opinion in Colloid & Interface Science. 2012;**17**(5):246-254

[7] de Oca-Ávalos JMM, Candal RJ, Herrera ML. Nanoemulsions: Stability and physical properties. Current Opinion in Food Science. 2017;**16**:1-6

[8] Troncoso E, Aguilera JM, McClements DJ. Fabrication, characterization and lipase digestibility of food-grade nanoemulsions. Food Hydrocolloid. 2012;**27**(2):355-363

[9] McClements DJ, Rao J. Food-grade nanoemulsions: Formulation, fabrication, properties, performance, biological fate, and potential toxicity. Critical Reviews in Food Science and Nutrition. 2011;**51**(4):285-330

[10] IUPAC. Compendium of Chemical Terminology (Version 2.3.3). Oxford: Gold Book Blackwell Scientific Publications; 1997. p. 499

[11] Mason TG, Wilking JN, Meleson K, Chang CB, Graves SM. Nanoemulsions: Formation, structure, and physical properties. Journal of Physics: Condensed Matter. 2006;**18**(41): R635-R666

[12] Walstra P. Principles of emulsion formation. Chemical Engineering Science. 1993;**48**(2): 333-349

[13] Tadros T, Izquierdo P, Esquena J, Solans C. Formation and stability of nano-emulsions. Advances in Colloid and Interface Science. 2004;**108-109**:303-318

[14] Wooster TJ, Golding M, Sanguansri P. Impact of oil type on nanoemulsion formation and Ostwald ripening stability. Langmuir. 2008;**24**(22):12758-12765

[15] Velikov KP, Pelan E. Colloidal delivery systems for micronutrients and nutraceuticals. Soft Matter. 2008;**4**(10):1964-1980

[16] Leong T, Wooster T, Kentish S, Ashokkumar M. Minimising oil droplet size using ultrasonic emulsification. Ultrasonics Sonochemistry. 2009;**16**(6):721-727

[17] Anton N, Benoit J-P, Saulnier P. Design and production of nanoparticles formulated from nano-emulsion templates—A review. Journal of Controlled Release. 2008;**128**(3):185-199

[18] Freitas S, Merkle HP, Gander B. Microencapsulation by solvent extraction/evaporation: Reviewing the state of the art of microsphere preparation process technology. Journal of Controlled Release. 2005;**102**(2):313-332

[19] Bouchemal K, Briançon S, Perrier E, Fessi H. Nano-emulsion formulation using spontaneous emulsification: Solvent, oil and surfactant optimisation. International Journal of Pharmaceutics. 2004;**280**(1-2):241-251

[20] Yin L-J, Chu B-S, Kobayashi I, Nakajima M. Performance of selected emulsifiers and their combinations in the preparation of β-carotene nanodispersions. Food Hydrocolloids. 2009;**23**(6):1617-1622

[21] Chu B-S, Ichikawa S, Kanafusa S, Nakajima M. Preparation and characterization of β-carotene nanodispersions prepared by solvent displacement technique. Journal of Agricultural and Food Chemistry. 2007;**55**(16):6754-6760

[22] Schubert H, Engel R. Product and formulation engineering of emulsions. Chemical Engineering Research and Design. 2004;**82**(9):1137-1143

[23] Jafari SM, Assadpoor E, He Y, Bhandari B. Re-coalescence of emulsion droplets during high-energy emulsification. Food Hydrocolloids. 2008;**22**(7):1191-1202

[24] Mohd-Setapar SH, Nian-Yian L, Kamarudin WNW, Idham Z, Norfahana A-T. Omega-3 emulsion of rubber (*Hevea brasiliensis*) seed oil. Agricultural Sciences. 2013;**4**(05):84

[25] Kentish S, Wooster T, Ashokkumar M, Balachandran S, Mawson R, Simons L. The use of ultrasonics for nanoemulsion preparation. Innovative Food Science & Emerging. 2008;**9**(2):170-175

[26] McClements DJ. Food Emulsions: Principles, Practices, and Techniques. United States: Taylor & Francis Group, CRC Press; 2015

[27] Walstra P. Physical Chemistry of Foods. United States: Taylor & Francis Group, CRC Press; 2003

[28] Quintanilla-Carvajal MX, Camacho-Díaz BH, Meraz-Torres LS, Chanona-Pérez JJ, Alamilla-Beltrán L, Jimenéz-Aparicio A, Gutiérrez-López GF. Nanoencapsulation: A new trend in food engineering processing. Food Engineering Reviews. 2010;**2**(1):39-50

[29] Sanguansri P, Augustin MA. Nanoscale materials development – A food industry perspective. Trends in Food Science & Technology. 2006;**17**(10):547-556

[30] Stang M, Schuchmann H, Schubert H. Emulsification in high-pressure homogenizers. Engineering in Life Sciences. 2001;**1**(4):151-157

[31] Schubert H, Ax K, Behrend O. Product engineering of dispersed systems. Trends in Food Science & Technology. 2003;**14**(1-2):9-16

[32] Wang X, Jiang Y, Wang Y-W, Huang M-T, Ho C-T, Huang Q. Enhancing anti-inflammation activity of curcumin through O/W nanoemulsions. Food Chemistry. 2008;**108**(2):419-424

[33] Yuan Y, Gao Y, Zhao J, Mao L. Characterization and stability evaluation of β-carotene nanoemulsions prepared by high pressure homogenization under various emulsifying conditions. Food Research International. 2008;**41**(1):61-68

[34] Ziani K, Chang Y, McLandsborough L, McClements DJ. Influence of surfactant charge on antimicrobial efficacy of surfactant-stabilized thyme oil nanoemulsions. Journal of Agricultural and Food Chemistry. 2011;**59**(11):6247-6255

[35] Buranasuksombat U, Kwon YJ, Turner M, Bhandari B. Influence of emulsion droplet size on antimicrobial properties. Food Science and Biotechnology. 2011;**20**(3):793-800

[36] Qian C, Decker EA, Xiao H, McClements DJ. Physical and chemical stability of β-carotene-enriched nanoemulsions: Influence of pH, ionic strength, temperature, and emulsifier type. Food Chemistry. 2012;**132**(3):1221-1229

[37] Salvia-Trujillo L, Rojas-Graü MA, Soliva-Fortuny R, Martín-Belloso O. Effect of processing parameters on physicochemical characteristics of microfluidized lemongrass essential oil-alginate nanoemulsions. Food Hydrocolloids. 2013;**30**(1):401-407

[38] Lin C-Y, Chen L-W. Comparison of fuel properties and emission characteristics of two-and three-phase emulsions prepared by ultrasonically vibrating and mechanically homogenizing emulsification methods. Fuel. 2008;**87**(10-11):2154-2161

[39] Abismaïl B, Canselier JP, Wilhelm AM, Delmas H, Gourdon C. Emulsification by ultrasound: Drop size distribution and stability. Ultrasonics Sonochemistry. 1999;**6**(1-2):75-83

[40] Maa Y-F, Hsu CC. Performance of sonication and microfluidization for liquid-liquid emulsification. Pharmaceutical Development and Technology. 1999;**4**(2):233-240

[41] Solè I, Maestro A, Pey C, González C, Solans C, Gutiérrez J. Nano-emulsions preparation by low energy methods in an ionic surfactant system. Colloids and Surfaces A: Physicochemical and Engineering Aspects. 2006;**288**(1-3):138-143

[42] Uson N, Garcia M, Solans C. Formation of water-in-oil (W/O) nano-emulsions in a water/mixed non-ionic surfactant/oil systems prepared by a low-energy emulsification method. Colloids and Surfaces A: Physicochemical and Engineering Aspects. 2004;**250**(1-3):415-421

[43] Izquierdo P, Esquena J, Tadros TF, Dederen C, Garcia M, Azemar N, Solans C. Formation and stability of nano-emulsions prepared using the phase inversion temperature method. Langmuir. 2002;**18**(1):26-30

[44] Morales D, Gutiérrez J, Garcia-Celma M, Solans Y. A study of the relation between bicontinuous microemulsions and oil/water nano-emulsion formation. Langmuir. 2003;**19**(18):7196-7200

[45] Horn D, Rieger J. Organic nanoparticles in the aqueous phase-theory, experiment, and use. Angewandte Chemie (International Ed. in English). 2001;**40**(23):4330-4361

[46] Shinoda K, Saito H. The effect of temperature on the phase equilibria and the types of dispersions of the ternary system composed of water, cyclohexane, and nonionic surfactant. Journal of Colloid and Interface Science. 1968;**26**(1):70-74

[47] Shinoda K, Saito H. The stability of O/W type emulsions as functions of temperature and the HLB of emulsifiers: The emulsification by PIT-method. Journal of Colloid and Interface Science. 1969;**30**(2):258-263

[48] Sadurní N, Solans C, Azemar N, García-Celma MJ. Studies on the formation of O/W nano-emulsions, by low-energy emulsification methods, suitable for pharmaceutical applications. European Journal of Pharmaceutical Sciences. 2005;**26**(5):438-445

[49] Israelachvili JN. Intermolecular and Surface Forces. United States: Elsevier, Academic Press; 2011

[50] Anton N, Gayet P, Benoit J-P, Saulnier P. Nano-emulsions and nanocapsules by the PIT method: An investigation on the role of the temperature cycling on the emulsion phase inversion. International Journal of Pharmaceutics. 2007;**344**(1-2):44-52

[51] Anton N, Vandamme TF. The universality of low-energy nano-emulsification. International Journal of Pharmaceutics. 2009;**377**(1-2):142-147

[52] Maestro A, Solè I, González C, Solans C, Gutiérrez JM. Influence of the phase behavior on the properties of ionic nanoemulsions prepared by the phase inversion composition method. Journal of Colloid and Interface Science. 2008;**327**(2):433-439

[53] Thakur RK, Villette C, Aubry J, Delaplace G. Dynamic emulsification and catastrophic phase inversion of lecithin-based emulsions. Colloids and Surfaces A: Physicochemical and Engineering Aspects. 2008;**315**(1-3):285-293

[54] Nedovic V, Kalusevic A, Manojlovic V, Levic S, Bugarski B. An overview of encapsulation technologies for food applications. Procedia Food Science. 2011;**1**:1806-1815

[55] Rijkers GT, de Vos WM, Brummer RJ, Morelli L, Corthier G, Marteau P. Health benefits and health claims of probiotics: Bridging science and marketing. The British Journal of Nutrition. 2011;**106**(9):1291-1296

[56] Hill C, Guarner F, Reid G, Gibson GR, Merenstein DJ, Pot B, Morelli L, Canani RB, Flint HJ, Salminen S, Calder PC, Sanders ME. Expert consensus document. The international scientific Association for Probiotics and Prebiotics consensus statement on the scope and appropriate use of the term probiotic. Nature Reviews. Gastroenterology & Hepatology. 2014;**11**(8):506-514

[57] Donsì F, Sessa M, Mediouni H, Mgaidi A, Ferrari G. Encapsulation of bioactive compounds in nanoemulsion-based delivery systems. Procedia Food Science. 2011;**1**:1666-1671

[58] Nelson KM, Dahlin JL, Bisson J, Graham J, Pauli GF, Walters MA. The essential medicinal chemistry of curcumin. Journal of Medicinal Chemistry. 2017;**60**(5):1620-1637

[59] Priyadarsini KI. The chemistry of curcumin: From extraction to therapeutic agent. Molecules. 2014;**19**(12):20091-20112

[60] Yu HL, Huang QR. Improving the oral bioavailability of curcumin using novel organo-gel-based nanoemulsions. Journal of Agricultural and Food Chemistry. 2012;**60**(21):5373-5379

[61] Huang Q, Yu H, Ru Q. Bioavailability and delivery of nutraceuticals using nanotechnology. Journal of Food Science. 2010;**75**(1):R50-R57

[62] Hatanaka J, Chikamori H, Sato H, Uchida S, Debari K, Onoue S, Yamada S. Physicochemical and pharmacological characterization of alpha-tocopherol-loaded nano-emulsion system. International Journal of Pharmaceutics. 2010;**396**(1-2):188-193

[63] Cheong JN, Tan CP, Man YBC, Misran M. α-Tocopherol nanodispersions: Preparation, characterization and stability evaluation. Journal of Food Engineering. 2008;**89**(2):204-209

[64] Fasolo D, Schwingel L, Holzschuh M, Bassani V, Teixeira H. Validation of an isocratic LC method for determination of quercetin and methylquercetin in topical nanoemulsions. Journal of Pharmaceutical and Biomedical Analysis. 2007;**44**(5):1174-1177

[65] Qian C, Decker EA, Xiao H, McClements DJ. Physical and chemical stability of beta-carotene-enriched nanoemulsions: Influence of pH, ionic strength, temperature, and emulsifier type. Food Chemistry. 2012;**132**(3):1221-1229

[66] Garti N, Spernath A, Aserin A, Lutz R. Nano-sized self-assemblies of nonionic surfactants as solubilization reservoirs and microreactors for food systems. Soft Matter. 2005;**1**(3):206

Chapter 4

Nanostructured Colloids in Food Science

Cristina Coman

Additional information is available at the end of the chapter

http://dx.doi.org/10.5772/intechopen.79882

Abstract

Nanostructured colloids are materials with at least one dimension in the nanometer range (<100 nm). Such materials find multiple and exciting applications in various areas of food science, and can lead to development of new and innovative food products and ingredients. Nanostructured colloids can be naturally present in food or they can be synthetically manufactured and added during different stages of food production and packaging. The building blocks of nanostructures in food consist of organic molecules (proteins, lipids, saccharides), inorganics (metal and metal oxides, carbon-based materials, clays) and combined organic and inorganic compounds. Some examples of nanostructured colloids naturally occurring in food include fat globules in homogenized milk, casein micelles, β-lactoglobulin fibers in milk. Synthetically manufactured colloids (artificial and engineered) include nanoemulsions, nanomicelles, nanocapsules, nanofoams, nanoliposomes, nanogels, nanofibers, metal and metal oxide nanoparticles. Synthetically manufactured nanostructures are normally added in food to enhance solubility, improve bioavailability, protect the biologically active compounds from degradation, increase the shelf life, color, flavor, and add nutritional value. Exciting fields of applications of nanostructured colloids in food science comprise: functional food ingredients, food additives, food supplements, food packaging and nanosensors.

Keywords: natural nanocolloids, artificial nanocolloids, engineered nanocolloids, applications, safety

1. Introduction

Nanotechnology provides innovative means of controlling and structuring food. Recently, nanotechnology has become an active research field in food science, especially related to the development of functional foods with improved functionality and value. Nanostructured colloids (nanocolloids) are nano-sized materials that can be inherently present in food, or they can be formed because of food processing technologies such as milling, homogenization, emulsification, electrospraying, spray-drying, supercritical CO_2-based techniques, gelation, foaming, etc. [1, 2].

According to the European Commission, "Nanomaterial" means a natural, incidental or manufactured material containing particles, in an unbound state or as an aggregate or as an agglomerate and where, for 50% or more of the particles in the number size distribution, one or more external dimensions is in the size range of 1–100 nm. In specific cases and where warranted by concerns for the environment, health, safety or competitiveness, the number size distribution threshold of 50% may be replaced by a threshold between 1 and 50% [3].

Additionally, engineered nanomaterial means any intentionally produced material that has one or more dimensions of the order of less than 100 nm, or that is composed of discrete functional parts, either internally or at the surface, many of which have one or more dimensions of the order of 100 nm or less, including structures, agglomerates or aggregates, which may have a size above the order of 100 nm but retain properties that are characteristic of the nanoscale. Properties characteristic of the nanoscale include: (i) those related to the large specific surface area of the materials considered; and/or (ii) specific physicochemical properties that are different from those of the non-nanoform of the same material [3].

Two approaches are commonly encountered in food science and industry to produce nano-sized particles: the so-called 'bottom-up' and 'top-down' approaches. The 'bottom-up' approach is related to the ability of molecules to self-assemble; it is a method based on atomic and molecular manipulation. Self-assembly is characteristic for the formation of naturally occurring nanostructures in foods. The 'top-down' approach leads to particle size reduction down to the nanometer range during physical processing of breaking-up bulk materials such as milling, homogenization, emulsification, and nanolithography [1, 4, 5].

Molecular self-assembly at the nanometer scale can be achieved by non-covalent interactions. The self-assembly of nanostructured food components is thermodynamically driven by non-covalent interactions such as van der Waals interactions, hydrogen bonds, electrostatic interactions, Coulomb interactions, coordinate bonding, or hydrophobic interactions. Proteins, peptides, lipids and polysaccharides have the ability to self-assemble into different types of nanostructures such as self-assembly of casein micelles, folding of globular proteins, formation of protein nanofibers, and formation of starch [6].

The 'top-down' approach most commonly includes dry milling, high-pressure homogenization, and ultrasound emulsification. In milling, mechanical energy is applied to break-up the particles and reduce their size to the nanometer range. Dry milling can be applied to reduce the size of wheat flour and increase thus its water-binding capacity. Also, applying dry milling to reduce the size of green tea particles resulted in increased oxygen-eliminating enzyme activity [5]. High-pressure homogenization or microfluidization converts high fluid pressure into shear forces, leading to uniform particle size reduction and formation of nanostructures. The technique is normally used in dairy sector for size reduction of fat globules which is useful to increase the stability of emulsions. Microfluidization results in size reduction and emulsion formation, leading to improved texture and aspect. It is applied for obtaining fillings and icings, salad dressings, yoghurts, syrups, flavored oil emulsions, creams, drinks [4, 5, 7]. Ultrasound emulsification is also used to prepare stable oil and water emulsions making use of high intensity ultrasound waves.

2. Types of nanostructured colloids in food science

2.1. Naturally occurring nanocolloids in food

In many cases, food products naturally contain ingredients in the nanometer range, which are different from the synthetically manufactured ones. There is a large variety of naturally occurring food structures that fall within the nanometer range at least in one dimension. Such nanostructures result from the self-assembly of the biological molecules through non-covalent interactions. The nanostructures can result from the arrangement of proteins, lipids, polysaccharides as-such, or in combinations, even combinations with small ligands. Also, most polysaccharides and lipids in food form linear polymeric chains less than 1 nm thick. Milk contains many naturally occurring nanostructures; during milk homogenization, lipid vesicles of around 100 nm are formed. Proteins in food are globular structures with size between 10 nm and hundreds of nm.

Some particular examples of commonly encountered protein nanostructures include: the casein molecule, β-lactoglobulin, bovine serum albumin, α-lactalbumin, lactoferrins (all present in milk), and lysozyme, ovalbumin, avidin (all present in egg white) [5, 8]. Most of the proteins of vegetal and animal origin are globular proteins, with few exceptions such as casein which forms micelles. Globular proteins such as whey proteins from milk have the ability to form particles with sizes of 40 nm.

Caseins belong to a family of phosphoprotein nanostructures present in milk that self-assemble into micellar granular structures which are suspended in the aqueous phase of milk. Ninety-five % of the caseins naturally self-assemble into micelles with 50–500 nm diameter [8, 9]. There are four varieties of caseins present in the mammalian milk: casein α_{s1}, α_{s2}, β, and κ [5, 10]. Casein accounts for 80% of the proteins in the cow's milk and 20–45% of the proteins in the human milk [11]. The casein micelle is stabilized by calcium-phosphate bonds. The high phosphate content of casein micelles allows it to associate with calcium and to form calcium phosphate salts. The high phosphate content of milk allows it to contain much more calcium than would be possible if all the calcium were dissolved in solution, thus casein proteins provide a good source of calcium for milk consumers. Thus, this type of colloidal micelles significantly increase the bioavailability of the calcium and phosphate ions [12]. Moreover, the casein micelles are important in milk digestion in the stomach and intestine, and constitute the basis for many of the milk products, being the major component of cheese (cheese is obtained by coagulation of casein). Casein is also used as food additive.

β-Lactoglobulin is the major whey protein in mammalian milk, except the human milk. The β-lactoglobulin monomer is a 3.6 nm long nanofiber which consists of a 162 amino acid sequence, and a molecular weight of 18.4 kDa. Depending on the pH, β-lactoglobulin can exist as slightly different structures such as dimmers, monomers, tetramers. The exact role of β-lactoglobulin has not yet been clearly established. It is believed to be involved in the transport of molecules due to its capacity to bind hydrophobic molecules and iron [5, 13].

α-lactalbumin is a 123 amino acid residue milk protein, with 2.01 nm radius, 14.2 kDa molecular weight and an isoelectric point between 4.2 and 4.5. α-Lactalbumin is a component of lactose-synthase, the enzyme responsible for lactose synthesis in mammalian milk, so it has a key role in regulating lactose production in milk. It has been proved that some folding variants of α-lactalbumin have bactericidal activity and some of them cause apoptosis of tumor cell [12, 14].

Lactoferrin is a 689 amino acid iron-binding protein with 3.6 nm radius, 82.4 kDa molecular weight and 8.7 isoelectric point. Lactoferrin has antibacterial and antifungal activity, it plays a role in iron adsorption, and studies have shown that lactoferrin is involved in the immune system responses [5, 15]. Human colostrum has the highest lactoferrin content, providing protective and antibacterial activity to infants. Lactoferrin normally complexes with other milk components such as casein, β-lactoglobulin, DNA, polysaccharides, and some of its biological activity is due its complexated forms [15].

Lysozyme (C-type) is a 129 amino acids globular antibacterial protein found in egg white, with a diameter of about 4.2 nm. C-type lysozyme is related to α-lactalbumin regarding the sequence structure [16].

Ovalbumin is the main protein in egg white, accounting for 55% of its protein content. It is a 385 amino acids sequence with 42.7 kDa molecular weight and 6.1 nm diameter [5, 17].

Lipid based nanocolloids naturally occurring in food are represented by lipoproteins in egg yolk, oleosomes found in plant seeds, fat globules in milk. The lipid nanostructures in food are normally composed of a hydrophobic core (triacylglycerols, glycerols, esterified fatty acids), surrounded by other phospholipids or proteins. The lipoproteins present in egg yolk form spherical structures with 15–60 nm diameter. The oleosomes are specialized lipid-based organelles in plant seeds, 0.1–10 μm in diameter, used for energy storage and for preventing fats from oxidation during germination. They are good vitamin reservoirs, being composed of a vitamin and triacylglycerol-rich core, surrounded by a phospholipid layer and protein layer [18].

2.2. Artificial nanocolloids in food

The term artificial refers to synthetically manufactured organic structures to produce colloidal particles on the nanoscale. There is huge interest towards designing nano- and microstructures for the food industry sector, since bioactive molecules can be encapsulated in such nanostructures. In many situations, encapsulation offers significant advantages, such as enhancing the stability and solubility of bioactive compounds (e.g. carotenoids), improving bioavailability (e.g. carotenoids, vitamins, minerals), protecting nutrients and bioactive compounds from degradation during manufacturing and storage, facilitating controlled release, masking unpleasant taste during eating (e.g. fish oil, polyphenols) [19].

Artificial organic colloidal nanostructures are most often synthesized from proteins, polysaccharides, and lipid molecules. Artificial organic nanostructures can be build-up of one type of molecule only, or, alternatively they can be combinations of different molecules. Commonly encountered artificial nanocolloids include nanoemulsions, nanomicelles, nanocapsules, nanofoams, nanoliposomes, nanogels, and nanofibers. The development of such structures has impact on the stability, protection, delivery, and bioavailability of bioactive molecules.

2.2.1. Nanoemulsions

Nanoemulsions are dispersions of at least two immiscible liquids, usually oil-in-water dispersions with mean diameters of 10–100 nm. Basically, they are fine dispersions of droplets of one liquid in another one in which the first is not soluble (usually defined as the oil phase and aqueous phase). Compared to conventional emulsions, nanoemulsions offer significant advantages for certain applications. First, nanoemulsions are much more stable compared to traditional emulsions, due to the small particle size [4, 20, 21]. They have higher stability towards particle aggregation and gravitational separation. In addition, nanoemulsions are used to increase solubility and bioavailability of bioactive hydrophobic molecules, especially carotenoids (lutein, lycopene, β-carotene), polyphenols (resveratrol), vitamins (A, D, E_3), enzymes (co-enzyme Q10), fatty acids (omega-3 fatty acids). Also, compared to traditional emulsions, the very small particle size of nanoemulsions, which can be smaller than the wavelength of light, only scatters light weakly. This does not alter the appearance of the food in which they are incorporated. So, they can be incorporated into optically transparent foods and beverages without affecting their clarity. Nanoemulsions can be prepared by high-pressure homogenization or ultrasound-assisted homogenization.

2.2.2. Nanoliposomes

Nanoliposomes are spherical bilayered vesicles made of amphiphilic molecules, such as phospholipids (e.g. lecithin, cholesterol). The molecules are arranged in two concentric circles, such that the hydrophilic end of the outer layer is exposed to the outer environment, while the inner hydrophilic end makes the hydrophilic core. The hydrophobic tails are in between the two hydrophilic layers. Their size varies between 20 and 400 nm usually. Liposomes can encapsulate both hydrophobic and hydrophilic molecules: hydrophobic molecules are encapsulated in the hydrophobic tail regions, while water soluble molecules are encapsulated in the hydrophilic core [4, 21].

2.2.3. Nanomicelles

Nanomicelles are spherical monolayered vesicles made of amphiphilic molecules. In a biological system, the molecules tend to arrange themselves in such a manner that the inner core is hydrophobic and the outer end is hydrophilic in nature. Thus, they can be used for encapsulation of hydrophobic molecules. Nanomicelles size varies in the range 5–100 nm.

2.2.4. Polymeric nanocapsules

Polymeric nanocapsules are synthetic colloidal nanostructures obtained from different natural or synthetic biocompatible polymers, with the final goal of encapsulating bioactive compounds. Most common polymers are poly lactic acid, poly-ε-caprolactone, poly-lactide-co-glycolide, natural polysaccharides (chitosan, Xanthan gum, Arabic gum), whey protein, etc. [22]. Nanocapsules are synthetic vesicles in which the bioactive compounds of interest, solubilized in an aqueous or oil core are covered by a polymeric shell. Polymeric nanocapsules can be build-up of single or multilayered polymeric walls, such as for example the polyelectrolyte multilayer capsules made of alternating layers of positively and negatively charged polymers. The bioactive compounds are normally encapsulated within the nanocapsule's core. Examples of molecules that can be encapsulated in polymeric nanostructures are the carotenoids lutein, lycopene, β-carotene, bixin, but also quercetin, α-tocopherol, vitamin B12, turmeric oil, lemongrass oil, cinnamon oil, etc. [22].

2.3. Engineered food nanocolloids

Engineered food nanocolloids mainly refer to synthetic metal and metal oxide nanoparticles. According to the European Commission, these nanoparticles are attracting great interest because they are increasingly used through sunscreen creams or other cosmetics, paints, plastics, dyes, food, medicines. Commonly used engineered colloidal systems in food industry include silver nanoparticles (AgNPs), titanium dioxide nanoparticles (TiO$_2$-NPs), zinc oxide nanoparticles (ZnO-NPs), silicon dioxide nanoparticles (SiO$_2$-NPs), nickel oxide nanoparticles (NiO-NPs), copper oxide nanoparticles (CuO-NPs), tin oxide nanoparticles (SnO$_2$-NPs), chromium oxide nanoparticles (Cr$_2$O$_3$-NPs), or composites between such nanoparticles [23]. A more complete list containing nanomaterials currently used in the European Union is presented in **Table 1**, based on an inventory of the European Food Safety Authority (EFSA) [3].

Naturally occurring nanomaterials	Nisin	Proteins	Casein
	Cellulose	Green tea	Lysozyme
	Starch	Enzymes	Zeolites
	Nanodroplets	Nanolipids	Nanosalts
Artificial nanomaterials	Nanocapsules (liposomes, micelles, nanocapsules, nanoemulsions)		
Engineered nanomaterials	Silver	Copper	Nickel
	Titanium dioxide	Fullerenes	Cerium oxide
	Nanocomposites	Selenium	Aluminum
	Zinc oxide	Calcium	Aluminum oxide
	Clays	Calcium carbonate	Carbon black
	Synthetic amorphous Silica	Calcium silicate	Organic Pigments
	Carbon nanotubes	Calcium phosphate	Platinum
	Silicon dioxide	Copper oxide	Sulfur
	Gold	Chromium	Amorphous Na-Al silicate
	Iron	Lead	Titanium nitride

Table 1. Nanomaterials in the agricultural, food and feed sector, as reported by an EFSA inventory [3].

Even if there is great potential for applications of such colloidal nanostructures in food indus-try, this class of structures is also related to potential food safety implications, health and environmental hazards. There is currently a continuous debate in this regard. The health hazards related to human exposure to such engineered nanostructures is due to the fact that their accumulation in tissues and cells is not entirely understood, and there is a fear of long term systemic toxicity. Environmental concerns are closely related to the fact that hundreds of tons of nanoparticles end up annually into the environment and further into the food chain. Plants play a critical role nanoparticles fate in the environment by assimilation, being exposed to nanoparticles due to agricultural soils fertilization with sewage sludge, as well as due to the use of nanoparticles in phytosanitary products.

3. Applications of nanostructured colloids in food science

The unique physical, chemical, and biological properties of colloidal nanostructures are consid-erably different from their bulk counterparts and this opens them unique and new possibilities for applications. Several research papers and reviews have identified potential applications of nanostructures for the food sector to improve food quality and safety. Increasing the shelf life of the food (preservation), pathogen detection, sensing, coloring, flavoring, and increasing nutri-tional value, are some of the important applications of nanostructures in the food and agri-food sectors.

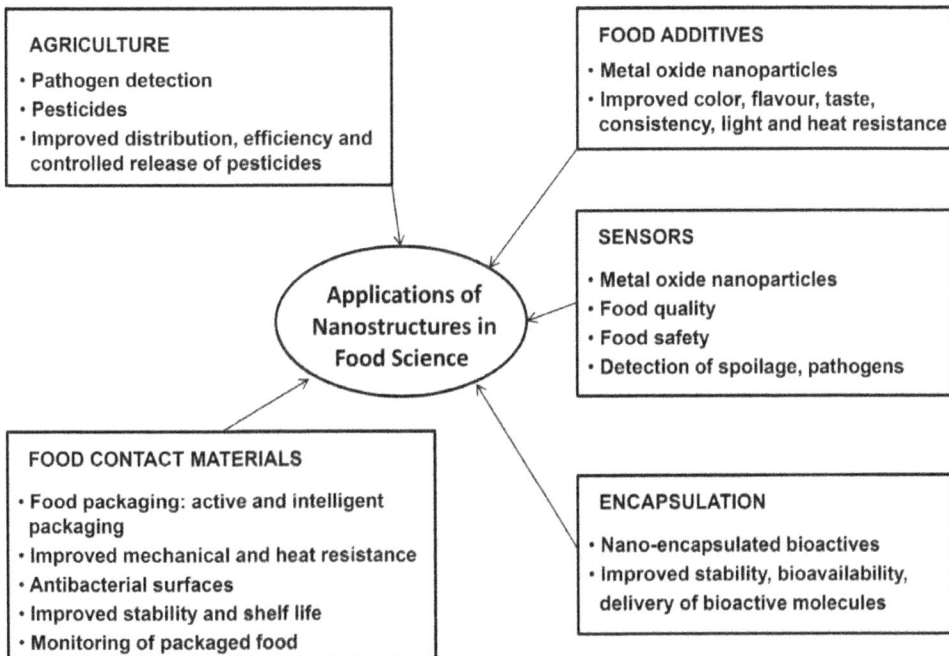

AGRICULTURE
- Pathogen detection
- Pesticides
- Improved distribution, efficiency and controlled release of pesticides

FOOD ADDITIVES
- Metal oxide nanoparticles
- Improved color, flavour, taste, consistency, light and heat resistance

Applications of Nanostructures in Food Science

SENSORS
- Metal oxide nanoparticles
- Food quality
- Food safety
- Detection of spoilage, pathogens

FOOD CONTACT MATERIALS
- Food packaging: active and intelligent packaging
- Improved mechanical and heat resistance
- Antibacterial surfaces
- Improved stability and shelf life
- Monitoring of packaged food

ENCAPSULATION
- Nano-encapsulated bioactives
- Improved stability, bioavailability, delivery of bioactive molecules

Figure 1. Applications of colloidal nanostructures in the food and agri-food sectors.

According to a report of EFSA [3], AgNPs and TiO_2-NPs are most used nanomaterials, with food additives and food contact materials being the most encountered applications. Some of the common potential fields of application of nanostructures and nanotechnology in food science and agriculture sectors are highlighted in **Figure 1**.

3.1. Sensing

Lately, there is an increasing number of sensing devices with applications in the food analysis sector, especially food quality control. Such devices are mainly based on colloidal metal oxide nanoparticles. They can mainly detect unwanted gases and volatile organic compounds that are produced in food products due to their spoilage. The working principle of such sensors is based on the interaction between the gas molecules and the colloidal metal oxide particles, which generates the change of a physical parameter in the transduction mechanism, making it possible to identify and in some situations to quantify the unwanted gas molecules. The colloidal metal oxide nanoparticle sensors can be independent devices used for food quality monitoring or they can be incorporated in intelligent packaging systems (see Section 3.3).

Some examples include sensors for the detection of trimethylamine, dimethylamine and ammonia, which are gas molecules naturally formed during biodegradation of plants, fish and animal tissues. Their presence is directly related to the degree of freshness of the fish and seafood products. Different metal oxide nanoparticle sensors have been reported for the detection of trimethylamine such as TiO_2-NPs, Au-WO_3-NPs, ZnO-Cr_2O_3-NPs, Cr_2O_3-SnO_2-NPs [23, 24]. For dimethylamine detection different formulations of ZnO-NPs have been reported. Besides being used for detection of fish meat freshness, detection of ammonia is also useful to give information about spoilage of other meat products.

3.2. Encapsulation and functional foods

Incorporation of several bioactive molecules in various food systems is limited by their poor water solubility and instability in presence of light, heat, and oxygen. As mentioned above, bioactive food compounds can be encapsulated by inclusion in different nano-sized structures, such as nanoemulsions, nanomicelles, nanoliposomes, nanocapsules. Vitamins, probiotics, fatty acids, lipids, antioxidants, preservatives, proteins, enzymes, peptides have been incorporated in several nanocapsule-based systems.

To summarize, the main objectives of encapsulation are:

- to enhance the solubility (e.g. coloring agents, antioxidants);
- to improve the bioavailability of the compounds, meaning the amount of bioactive compound which is absorbed by the human body (antioxidants, vitamins, minerals, enzymes);
- to improve stability and shelf life;
- to protect bioactive molecules and micronutrients during manufacture, storage, retail;
- to allow controlled release of bioactive molecules and drugs (only at desired target organs, only in particular parts of the gastrointestinal tract; e.g. pH-driven release, light-driven release).

As particular example, enzymes are a very important class of bioactive molecules for which colloidal nanoencapsulation can bring advances regarding improvement of their stability, activity, avoiding of denaturation, improving absorption in the gastro-intestinal tract. Enzymes would benefit delivery through functional foods and supplements, but since they are susceptible to denaturation, encapsulation is one was to overcome the problem. Food-grade colloidal enzyme delivery nanocapsules are normally composed of lipids, proteins, polysaccharides (starch, carrageenan, etc.), and encapsulating structures include colloidal nanoemulsions, solid-lipid nanoparticles, liposomes, gels [25].

Micelles and liposomes are used as encapsulating and carrier structures for different hydrophobic molecules such as essential oils, flavors, antioxidants (polyphenols, carotenoids, coenzyme Q10), vitamins, minerals, proteins, nutrients, nutraceuticals [4, 21].

Nanoemulsions find applications in the production of table spreads and yoghurts, flavored oils, personalized beverages, sweeteners, fortification of milk with vitamins, minerals, antioxidants [4, 5].

Nanoliposomes in food were reported as vesicles for encapsulation of antioxidants, fatty acids, oils. Stable nanoliposomes made of ω-3 and ω-6 fatty acids, tocopherol and vitamin C encapsulated in soy phosphatidylcholine were studied as functional ingredients in acidic foods, such as the orange juice [26].

Nanocapsules can be incorporated in interactive foods and drinks which can release flavors, aromas, colors by breaking up at certain pH values or at certain infrared frequencies. Breaking up causes delivery and release of the content. One such example of colloidal polymeric nanocapsules is the light-responsive polyelectrolyte capsules containing gold nanoparticles as light responsive nanomaterial [27]. These colloidal nanocapsules are formed by alternating layers of positively and negatively charged, biocompatible polymers, hold together by electrostatic interactions. The bioactive molecules are encapsulated into the nanocapsules core, while the optically addressable colloidal AuNPs are embedded within the nanocapsule walls. The AuNPs within the capsule walls absorb light in the near infrared region, where most tissues show only weak absorption; the laser energy is then efficiently transformed into heat, being possible to open the capsules and achieve laser-induced release of the encapsulated molecules, with minimal damage of the surrounding tissues and of the bioactive molecules. There is currently a lot of effort being put into developing this type of light-responsive systems. One potential application could be the encapsulation of antioxidants such as carotenoids and their efficient delivery to retina cells—to be used as food supplements and to prevent and treat age-dependent macular degeneration.

3.3. Food packaging and antimicrobials

Lately, several innovative food packaging materials based on colloidal nanoparticles have been developed. Nano-packaging is a new generation of food packaging technology which represents a radical alternative to the conventional food packaging. Basically, nanomaterials are incorporated in the polymer matrix to offer mechanical strength or to function as a barrier against gases, volatile components (e.g. flavors) or moisture. Widely used colloidal nanostructures in food packaging include colloidal AgNPs, ZnO-NPs, TiO$_2$-NPs, mostly used for antimicrobial activity and ultraviolet protection, also nanoclays used to develop materials with enhanced gas-barrier properties, colloidal Silica nanoparticles used for surface coating of packaging materials. Titanium nitride in the nanoparticle form is authorized in the EU for use as additive or polymer production aid in plastic food contact materials (EU Regulation 1183/2012) [28].

The concept of active packaging, e.g. antimicrobial packaging involving incorporation of nanoparticles with antibacterial and antioxidant properties (AgNPs, ZnO-NPs, MgO-NPs, TiO_2-NPs) into the food package has attracted increasing interest due to its potential huge impact in the food safety sector [29]. An active packaging system involves interaction between the packaging material and the food to provide desirable effects, such as microbial safety, extended shelf life of foods. For example, colloidal AgNPs have been for long known to be effective antibacterials. They are reported by several authors [30, 31] to be incorporated in antibacterial coatings, efficient against different fungi, gram positive and gram-negative pathogens such as *Escherichia coli, Staphylococcus aureus, Aspergillus niger, Penicillium funiculosum, Chaetomium globosum, Aspergillus terreus*, and *Aureobasidium pullulans*.

Other nanocolloidal systems (e.g. TiO_2-NPs, SiO_2-NPs, nanoclays) can ensure good food preservation by blocking the UV radiation, improving mechanical and heat-resistance properties, reducing the permeability of foils, deodorizing, antimicrobials [32]. Colloidal TiO_2-NPs are UV blocking agents, used as filler particles in foils, food packaging, plastic containers. Nanoclays are among the first polymer nanostructures to emerge on the market as improved materials for food packaging [33]. Clays are one of the oldest and most important types of available colloidal materials. Nanoclay based food packaging increases the shelf life of oxygen sensitive foods by increasing the water and gas barrier of the polymer material. Uses have been reported for manufacturing of bottles for beers and carbonated drinks [33]. Nanoclays have been also reported to improve mechanical properties, thermal stability and fire resistance of polyethylene, polypropylene, nylon, poly(e-caprolactone), polyethylene terephthalate polymers.

The possibility to improve the performances of polymers for food packaging by adding nanoparticles has led to the development of a variety of polymer nanomaterials. Polymers with nanofillers, e.g. polymer nanocomposites, are created by dispersing an inert, nanofiller into a polymeric matrix. Filler materials can include nanoclays as mentioned above, SiO_2-NPs, carbon nanotubes, starch nanocrystals, cellulose-based nanofibers, chitin and chitosan nanoparticles [29, 32]. Compared to conventional polymers, polymer nanocomposites show improved packaging properties by having better mechanical resistance, flame resistance, and better thermal properties (e.g. melting points, degradation and glass transition temperatures). Polymer nanocomposites can improve quality of meat and meat products by reducing moisture loss, reducing lipid peroxidation, improving thus the appearance of the products by maintaining the flavors, color, and texture.

Very recent work in the field of nanocomposites and antimicrobial packaging reports on the use of polyvinyl alcohol/nanocellulose/Ag nanocomposite films with antibacterial activity against *Staphylococcus aureus* and *Escherichia coli* [34]. Other advances report the development of nanocomposite gelatin based antifungal films efficient against *Aspergillus niger* [35, 36]. For the gelatin-based nanocomposite preparation either chitin nanofibers and ZnO-NPs, or chitin nanofibers and corn oil were used.

Intelligent packaging involves incorporation of nanosensors into food packaging to improve detection and tracking of the physicochemical changes in food during storage and transport. It gives information based on its ability to detect changes in the product's environment [29].

Despite tremendous advances in applying nanostructures in food packaging, there is continuous debate regarding the potential diffusion in time of nano-sized materials from food packaging into the packaged food and this is a topic that requires further and careful investigation [37].

3.4. Food additives

3.4.1. Titanium dioxide (TiO$_2$)

TiO$_2$ is a food coloring agent approved by EFSA, known as E171. It is used as whitener and colorant. It is also used as a food additive and flavor enhancer in a variety of non-white foods, including dried vegetables, nuts, seeds, soups, and mustard, as well as beer and wine. E171 contains TiO$_2$ particles, partially in the nanometer size. The average particle size is 200–300 nm, part of this bulk material may contain a fraction of particles with sizes <100 nm. A recent study showed that 5–36% of the TiO$_2$ in food products is in the nano-size range [38]. Colloidal TiO$_2$ can be found in products such as candies, sweets, chewing gum, toothpaste, nutritional supplements, sunscreen protection products [39].

Other metals in the form of colloidal nano-sized particles are available as food or health supplements. These include selenium nanoparticles [40], calcium nanoparticles, iron nanoparticles, and colloidal suspensions of metal particles, e.g. cobalt nanoparticles, gold nanoparticles, platinum nanoparticles, silver nanoparticles, molybdenum nanoparticles, paladium nanoparticles, titanium nanoparticles, and zinc nanoparticles.

Extensive use of TiO$_2$ has been, in some studies linked with adverse health effects. The European Chemicals Agency has concluded that titanium dioxide may cause cancer if inhaled [41]. In their study on mice, Rizk et al. [42] found that some biochemical parameters and the liver structure were influenced by colloidal TiO$_2$-NPs in a dose-dependent manner. There is continuous debate and further studies are needed to elucidate all the safety aspects of TiO$_2$ in its nano- and non-nanoforms on pharmaceutical and food applications. In September 2016, EFSA published an opinion on the re-evaluation of E171, based on a detailed literature review on TiO$_2$ nanoparticles. It was concluded that current exposure of consumers to E171 related to its use in foodstuffs is not likely to constitute a health risk, but that it was not possible to establish an acceptable daily intake [43].

3.4.2. Zinc oxide (ZnO)

Zinc oxide nanoparticles (ZnO-NPs) are used in some food contact materials such as polypropylene and polyethylene. ZnO is used as a transparent ultraviolet light absorber in unplasticized polymers at up to 2% by weight [44]. It is added as nanopowder in the formulations and thus colloidal nanoparticles are present in the final polymer. It absorbs UV light without re-emitting it as heat, improving thus the stability of the polymers. It is also used in nutritional supplements, for example vitamins [39].

There are several *in vitro* studies demonstrating some degree of toxicity of ZnO-NPs. A study on RKO human colon carcinoma cells found that the toxicity is not related to the colloidal ZnO-NPs concentration, but to the direct particle-cell contact [45]. *In vivo* studies have shown that ZnO in the nanometer form tends to accumulate in liver, spleen, kidney tissues in higher amounts compared to the micrometer ZnO particles. EFSA has performed a safety assessment study of ZnO-NPs used in food contact materials [44]. Considering previous knowledge on the nanoparticles diffusion in polymers and the solubility characteristics of the ZnO-NPs, it was concluded that ZnO-NPs do not migrate and the focus should be on the migration of soluble ionic Zn, which complies with the current specific migration limit.

3.4.3. Silicon dioxide (SiO$_2$)

Silicon dioxide (SiO$_2$) is a licensed food additive (E551), used as anticaking agent. It is one of the most important caking agents, present in many powdered food items, chewing gums, cheese, seasonings, cooking salt [3]. It is also used for clearing of beverages. E551contains primary particles, aggregates and agglomerates, and partially it contains SiO$_2$ in the nanometer range, with sizes <100 nm.

3.5. Agriculture and animal feed

Nanostructured colloidal materials have been reported to have contributions in the delivery of agro-chemicals such as pesticides and herbicides through, for example controlled delivery systems based on encapsulation. Such systems can decrease the amounts of sprayed chemicals by more efficient delivery of bioactive compounds. Different nanostructured delivery systems could be applied to store, protect, deliver, release of pesticides, nutrients, fertilizers [5]. Some formulations of nanocapsules with pesticides and herbicides have been reported. Some companies have produced nanosuspensions and nanoemulsions containing water or oil soluble pesticides and herbicides. Nanocapsules can also be used for the delivery of DNA and chemicals to plant tissues, with the final goal of ensuring protection against several pests and diseases [5].

Additionally, uses of colloidal nanostructures in agriculture have been reported for wastewater treatment, disinfectants (AgNPs mainly), sensors for detection of pesticides, fertilizers, herbicides, pathogens, aflatoxins, soil pH and moisture, etc. The efficiency of colloidal AgNPs and colloidal gold nanoparticles (AuNPs) as larvicide against *Aedes aegypti* was reported [46]. AgNPs were also found to be effective against powdery mildew in cucumber and pumpkin, under different cultivation conditions *in vitro* and *in vivo* [47]. Powdery mildew is one of the most devastating diseases in cucurbits, which can lead to serious crop yield decrease. Colloidal AgNPs have been studied also as an alternative for the antibiotics used in the poultry production [48].

Colloidal nano-sized minerals, vitamin or other additives could equally be used for animal feed, to ensure improved delivery of veterinary drugs and to improve availability and quality of nutrients in feed.

3.6. Some examples of commercial nanofood products

Many food companies are investigating or already employing nanocolloids to change the structure of food and drinks. Examples include the interactive foods and drinks containing colloidal nanocapsules that can change color and flavor, spreads and ice creams with nanopar-ticle emulsions that improve food texture [49]. In the following, some examples of commercial nanofood products existing on the market will be provided.

Nanotea [50] is a colloidal formulation containing nano-selenium in concentration of 3–5 ppm, prepared using nanotechnology. The nanotea can release effectively all the excellent essences of the tea, being a good selenium supplement (10-fold increase).

Nanoceuticals™ Slim Shake Chocolate [51] is a slimming product containing Cocoa nanoclusters reported to enhance the taste benefits of the product.

Canola Active Oil [52] produced by Shemen Industries, Israel, is a product obtained by the nano-sized self-assembled structured liquids (NSSL) technique, giving compressed micelles, called nanodrops. These micelles act as liquid carriers, allowing encapsulation of different bioactive components (vitamins, minerals, phytosterols). The micelles are added to the food products for improving bioavailability of the bioactives, being able to pass through the diges-tive system effectively, without breaking up, and thus effectively reaching the absorption site.

Hydracel [53] is a food supplement containing nanocolloidal minerals, made of water, silicon, magnesium sulfate, potassium carbonate, potassium hydroxide, sunflower oil, reported to improve the cell life cycle.

Nano calcium [54] is a food supplement containing calcium in the nanoparticle colloidal form, reported to enhance and fasten calcium absorption in the body.

Production of carotenoid preparations in the form of coldwater-dispersible nano powders with up to 200 nm size, and the use of the novel carotenoid preparations are reported by US patent 5968251A [55].

LycoVit [56] is a BASF product containing lycopene in the nanosize range. Its intended use is for dietary supplement, beverage and food applications such as turbid beverages, carbonated drinks, cake and biscuit fillings, dairy desserts, candies, pastas, meats and soups. It is reported to be a safe, highly effective substitute for artificial colors, which may have an adverse effect on activity and attention in children.

NovaSOL curcumin [57] is a nanomicelle-based compound, stable through the digestive pro-cess, able to deliver curcumin to the intestinal wall with maximum therapeutic effect. It is reported that NovaSOL curcumin has 185× increased bioavailability than native powdered turmeric/curcumin.

NutraLease Ltd. Company [58], through the nano-sized self-assembled structured liquids (NSSL) technology develops nanocarriers to be incorporated in food systems and cosmetics. The tech-nology allows enhanced solubility and bioavailability of compounds in water-based or oil-based media. The encapsulated compounds include lycopene, beta-carotene, lutein, phytosterols, coenzyme Q10, lipoic acid, and DHA/EPA. The company focuses on fortifying foods and beverages.

4. Regulatory aspects of nanotechnology in the agri-food sector in the EU countries

As mentioned, nanotechnology and the use of colloidal nanoparticles in the food science and agricultural sectors open up the possibility to produce new ingredients and foods with improved and beneficial properties. However, as the research in the field is intensifying, in some cases several health and environmental concerns appear. Therefore, a careful assessment of the potential risks is essential before approval of any nanostructured ingredients. Efforts are being made worldwide to ensure the production and safe use and handling of nanomaterials in the agricultural and food sector.

Nanomaterials are present in many commercial preparations. Some of the nanoparticles are ingested by humans through food consumption. EFSA observed that organic artificial nanomaterials (vesicles, nanoemulsions) present a low risk for health as they are most likely completely metabolized. In contrast, the agency considered that the research effort has to be concentrated on inorganic, synthetic nanomaterials, whose faith and accumulation in the human body is not exactly certain, nor are their effects upon long term exposure.

There is no single legislation fully dedicated to nanomaterials, but there are several regulations addressing different aspects of nanomaterials use in the agri-food sector. **Table 2** summarizes relevant European regulations connected to the use of nanomaterials in food. The "General Principles and Requirements of Food Law" are established by the EC Regulation 178/2002, containing several articles on novel foods, novel foods ingredients, food and feed additives,

Legislation number	Legislation topic
(EC) 178/2002	General principles and requirements of food law
(EU) 2283/2015	Novel foods, novel food ingredients, food and feed additives, food supplements, vitamins, minerals, food contact materials
(EC) 1332/2008	Food enzymes
(EC) 1333/2008	Food additives
(EC) 1334/2008	Food flavorings
(EC) 46/2002	Food supplements
(EC) 10/2011	Plastic materials and food contact materials
(EC) 450/2009	Active and intelligent materials and food contact materials
(EU) 528/2012	Biocides
(EC) 1107/2009	Plant protection products

Table 2. Overview of the European legislation on the agri-food sector, in relation to nanomaterials.

food supplements, vitamins, minerals and food contact materials. Regulation (EU) 2283/2015 concerning novel foods and novel food ingredients covers aspects of the use of nanotechnology in food production and repeals the older Regulation (EC) 258/97. Food additives are covered by several regulations: Regulation (EC) 1333/2008 on food additives, Regulation (EC) 1332/2008 on food enzymes, Regulation (EC) 1334/2008 on flavorings and certain food ingredients with flavoring properties for use in food. Regulation 1331/2008 establishes a common authorization procedure for food additives, enzymes and flavorings prior to their entrance on the market. Vitamins and minerals are regulated by Directive (EC) 46/2002 on food supplements. Regulation (EC) 10/2011 establishes specific rules for plastic materials and articles intended to come in contact with food, to be applied for their safe use. Regulation (EC) 450/2009 deals with active and intelligent materials and articles intended to come into contact with food. The Biocidal Products Regulation (EU) 528/2012 lays down the provisions for the use of non-agricultural pesticides by both professionals and consumers and Regulation (EC) 1107/2009 concerns placing of plant protection products on the market [3, 59, 60].

5. Conclusions

Applications of colloidal nanostructures in food science are receiving increased attention due to the possibility of developing structures and materials with improved properties compared to the conventional ones. In the future, enabling a better control over the colloidal nanostructure formation will enhance the control over food structure formation, leading to the design of new food products with improved characteristics and enhanced impact on consumers'. It is the unique physical, chemical, optical and biological properties of the nanocolloidal systems, considerably different from those of the corresponding bulk materials that are responsible for their unique properties. Colloidal nanostructures are also raising a series of health and environmental concerns, since not so much information is known on the long-term exposure of organisms to such materials, even if at the present time most of them are considered safe. In the future, more studies and regulations regarding the health and environmental impact of colloidal nanostructures will be needed.

Despite the fate and potential toxicity of such colloidal nanostructures being not fully understood, it is obvious that this research field will bring significant advances in the food sector, which will most likely impact the food safety and nutrition sectors, the production of new functional foods and ingredients, the developing of innovative packaging with great potential to transform the future of food packaging, will as well help assisting in the detection of pesticides, pathogens, toxins, etc.

Acknowledgements

This work was supported by a grant of Ministry of Research and Innovation, CNCS—UEFISCDI, project number PN-III-P1-1.1-TE-2016-1907, within PNCDI III.

Conflict of interest

The authors declare there is no conflict of interest.

Author details

Cristina Coman

Address all correspondence to: cristina.coman@usamvcluj.ro

University of Agricultural Sciences and Veterinary Medicine Cluj-Napoca, Cluj-Napoca, Romania

References

[1] Huang Q, editor. Nanotechnology in the Food, Beverage and Nutraceutical Industries. Cambridge: Woodhead Publishing in Food Science Technology and Nutrition; 2012. DOI: 10.1533/9780857095657

[2] Dickinson E, van Vliet T, editors. Food Colloids, Biopolymers and Materials. Cambridge: The Royal Society of Chemistry; 2003. DOI: 10.1016/s0927-7765(03)00133-4

[3] Inventory of Nanotechnology applications in the agricultural, feed and food sector [Internet]. 2014. Available from: https://efsa.onlinelibrary.wiley.com/doi/epdf/10.2903/sp.efsa.2014.EN-621 [Accessed: 2018-05-08]

[4] Pathakoti K, Manubolu M, Hwang HM. Nanostructures: Current uses and future applications in food science. Journal of Food and Drug Analysis. 2017;25(2):245-253. DOI: 10.1016/j.jfda.2017.02.004

[5] Axelos MAV, Van de Voorde M, editors. Nanotechnology in Agriculture and Food Science. Weinheim: Wiley-VCH; 2017. DOI: 10.1002/9783527697724

[6] Whitesides GM, Boncheva M. Beyond molecules: Self-assembly of mesoscopic and macroscopic components. Proceedings of the National Academy of Sciences of the United States of America. 2002;99(8):4769-4774. DOI: 10.1073/pnas.082065899

[7] Augustin MA, Sanguansri P. Nanostructured materials in the food industry. Advances in Food and Nutrition Research. 2009;58:183-213. DOI: 10.1016/S1043-4526(09)58005-9

[8] Rogers MA. Naturally occurring nanoparticles in food. Current Opinion in Food Science. 2016;7:14-19. DOI: 10.1016/j.cofs.2015.08.005

[9] Holt C, Horne DS. The hairy casein micelle: Evolution of the concept and its implications for dairy technology. Netherlands Milk and Dairy Journal. 1996;50(2):85-111

[10] Dalgleish DG. Casein micelles as colloids: Surface structures and stabilities. Journal of Dairy Science. 1998;**81**(11):3013-3018. DOI: 10.3168/jds.S0022-0302(98)75865-5

[11] Kunz C, Lonnerdal B. Human-milk proteins—Analysis of casein subunits by anion-exchange chromatography. Gel-Electrophoresis, and Specific Stainign Methods. American Journal of Clinical Nutrition. 1990;**51**(1):37-46

[12] Milk Composition [Internet]. Available from: http://www.milkfacts.info/Milk%20Composition/Protein.htm#MilkProtPhysProp [Accessed: 2018-03-15]

[13] Brownlow S, Cabral JHM, Cooper R, Flower DR, Yewdall SJ, Polikarpov I, et al. Bovine beta-lactoglobulin at 1.8 angstrom resolution—Still an enigmatic lipocalin. Structure. 1997;**5**(4):481-495. DOI: 10.1016/s0969-2126(97)00205-0

[14] Permyakov EA, Berliner LJ. Alpha-lactalbumin: Structure and function. FEBS Letters. 2000;**473**(3):269-274. DOI: 10.1016/s0014-5793(00)01546-5

[15] Sanchez-Valdes S, Ortega-Ortiz H, Valle L, Medellin-Rodriguez FJ, Guedea-Miranda R. Mechanical and antimicrobial properties of multilayer films with a polyethylene/silver nanocomposite layer. Journal of Applied Polymer Science. 2009;**111**(2):953-962. DOI: 10.1002/app.29051

[16] Peters CWB, Kruse U, Pollwein R, Grzeschik KH, Sippel AE. The human lysozyme gene-sequence organization and chromosomal localization. European Journal of Biochemistry. 1989;**182**(3):507-516. DOI: 10.1111/j.1432-1033.1989.tb14857.x

[17] Nisbet AD, Saundry RH, Moir AJG, Fothergill LA, Fothergill JE. The complete amino acid sequence of hen ovalbumin. European Journal of Biochemistry. 1981;**115**(2):335-345. DOI: 10.1111/j.1432-1033.1981.tb05243.x

[18] Stumpf PK, Mudd JB, Nes WD. The Metabolism, Structure, and Function of Plant Lipids. New York: Springer; 1987. DOI: 10.1007/978-1-4684-5263-1

[19] Nedovic V, Kalusevic A, Manojlovic V, Levic S, Bugarski B. An overview of encapsulation technologies for food applications. In: Saravacos G, Taoukis P, Krokida M, Karathanos V, Lazarides H, Stoforos N, et al., editors. 11th International Congress on Engineering and Food. Amsterdam: Procedia Food Science; 2011. pp. 1806-1815. DOI: 10.1016/j.profoo.2011.09.266

[20] Choi AJ, Kim CJ, Cho YJ, Hwang JK, Kim CT. Characterization of capsaicin-loaded nanoemulsions stabilized with alginate and chitosan by self-assembly. Food and Bioprocess Technology. 2011;**4**(6):1119-1126. DOI: 10.1007/s11947-011-0568-9

[21] Greßler S, Gazsó A, Simkó M, Nentwich M, Fiedeler U. Nanoparticles and nanostructured materials in the food industry. NanoTrust Dossiers. 2010;**4**:1-7. DOI: 10.1016/S1043-4526(09)58005-9

[22] dos Santos PP, Flores SH, Rios AD, Chiste RC. Biodegradable polymers as wall materials to the synthesis of bioactive compound nanocapsules. Trends in Food Science & Technology. 2016;**53**:23-33. DOI: 10.1016/j.tifs.2016.05.005

[23] Galstyan V, Bhandari MP, Sberveglieri V, Sberveglieri G, Comini E. Metal oxide nanostructures in food applications: Quality control and packaging. Chemosensors. 2018;6(2):1-21. DOI: 10.3390/chemosensors6020016

[24] Perillo PM, Rodriguez DF. Low temperature trimethylamine flexible gas sensor based on TiO$_2$ membrane nanotubes. Journal of Alloys and Compounds. 2016;657:765-769. DOI: 10.1016/j.jallcom.2015.10.167

[25] McClements DJ. Encapsulation, protection, and delivery of bioactive proteins and peptides using nanoparticle and microparticle systems: A review. Advances in Colloid and Interface Science. 2018;253:1-22. DOI: 10.1016/j.cis.2018.02.002

[26] Marsanasco M, Piotrkowski B, Calabro V, Alonso SD, Chiaramoni NS. Bioactive constituents in liposomes incorporated in orange juice as new functional food: Thermal stability, rheological and organoleptic properties. Journal of Food Science and Technology-Mysore. 2015;52(12):7828-7838. DOI: 10.1007/s13197-015-1924-y

[27] Sukhorukov GB, Volodkin DV, Gunther AM, Petrov AI, Shenoy DB, Mohwald H. Porous calcium carbonate microparticles as templates for encapsulation of bioactive compounds. Journal of Materials Chemistry. 2004;14(14):2073-2081. DOI: 10.1039/b402617a

[28] Commission Regulation (EU) No 1183/2012 [Internet]. 2012. Available from: https://eur-lex.europa.eu/legal-content/EN/TXT/?uri=celex%3A32012R1183 [Accessed: 2018-04-28]

[29] Huang JY, Li X, Zhou WB. Safety assessment of nanocomposite for food packaging application. Trends in Food Science & Technology. 2015;45(2):187-199. DOI: 10.1016/j.tifs.2015.07.002

[30] Gunell M, Haapanen J, Brobbey KJ, Saarinen JJ, Toivakka M, Makela JM, et al. Antimicrobial characterization of silver nanoparticle-coated surfaces by "touch test" method. Nanotechnology Science and Applications. 2017;10:137-145. DOI: 10.2147/nsa.S139505

[31] Chen KH, Ye WJ, Cai SL, Huang L, Zhong TS, Chen LK, et al. Green antimicrobial coating based on quaternised chitosan/organic montmorillonite/Ag NPs nanocomposites. Journal of Experimental Nanoscience. 2016;11(17):1360-1371. DOI: 10.1080/17458080.2016.1227095

[32] Duncan TV. Applications of nanotechnology in food packaging and food safety: Barrier materials, antimicrobials and sensors. Journal of Colloid and Interface Science. 2011;363(1):1-24. DOI: 10.1016/j.jcis.2011.07.017

[33] Silvestre C, Duraccio D, Cimmino S. Food packaging based on polymer nanomaterials. Progress in Polymer Science. 2011;36(12):1766-1782. DOI: 10.1016/j.progpolymsci.2011.02.003

[34] Sarwar MS, Niazi MBK, Jahan Z, Ahmad T, Hussain A. Preparation and characterization of PVA/nanocellulose/Ag nanocomposite films for antimicrobial food packaging. Carbohydrate Polymers. 2018;184:453-464. DOI: 10.1016/j.carbpol.2017.12.068

[35] Sahraee S, Ghanbarzadeh B, Milani JM, Hamishehkar H. Development of gelatin bionanocomposite films containing chitin and ZnO nanoparticles. Food and Bioprocess Technology. 2017;10(8):1441-1453. DOI: 10.1007/s11947-017-1907-2

[36] Sahraee S, Milani JM, Ghanbarzadeh B, Hamishehkar H. Effect of corn oil on physical, thermal, and antifungal properties of gelatin-based nanocomposite films containing nano chitin. LWT—Food Science and Technology. 2017;**76**:33-39. DOI: 10.1016/j.lwt.2016.10.028

[37] Mackevica A, Olsson ME, Hansen SF. Silver nanoparticle release from commercially available plastic food containers into food simulants. Journal of Nanoparticle Research. 2016;**18**(1):1-11. DOI: 10.1007/s11051-015-3313-x

[38] Weir A, Westerhoff P, Fabricius L, Hristovski K, von Goetz N. Titanium dioxide nanoparticles in food and personal care products. Environmental Science & Technology. 2012;**46**(4):2242-2250. DOI: 10.1021/es204168d

[39] Martirosyan A, Schneider YJ. Engineered nanomaterials in food: Implications for food safety and consumer health. International Journal of Environmental Research and Public Health. 2014;**11**(6):5720-5750. DOI: 10.3390/ijerph110605720

[40] Xu J, Yang FM, An XX, Hu QH. Anticarcinogenic activity of selenium-enriched green tea extracts in vivo. Journal of Agricultural and Food Chemistry. 2007;**55**(13):5349-5353. DOI: 10.1021/jf070568s

[41] European chemicals body links titanium dioxide to cancer [Internet]. 2017. Available from: https://www.chemistryworld.com/news/european-chemicals-body-links-titanium-dioxide-to-cancer/3007557.article [Accessed: 2018-05-23]

[42] Rizk MZ, Ali SA, Hamed MA, El-Rigal NS, Aly HF, Salah HH. Toxicity of titanium dioxide nanoparticles: Effect of dose and time on biochemical disturbance, oxidative stress and genotoxicity in mice. Biomedicine & Pharmacotherapy. 2017;**90**:466-472. DOI: 10.1016/j.biopha.2017.03.089

[43] Aguilar F, Crebelli R, Di Domenico A, Dusemund B, Frutos MJ, Galtier P, et al. Re-evaluation of titanium dioxide (E 171) as a food additive. EFSA Journal. 2016;**14**(9):4545. DOI: 10.2903/j.efsa.2016.4545

[44] Bolognesi C, Castle L, Cravedi JP, Engel KH, Franz R, Fowler P, et al. Safety assessment of the substance zinc oxide, nanoparticles, for use in food contact materials. EFSA Journal. 2016;**14**(3):4408. DOI: 10.2903/j.efsa.2016.4408

[45] Moos PJ, Chung K, Woessner D, Honeggar M, Cutler NS, Veranth JM. ZnO particulate matter requires cell contact for toxicity in human colon cancer cells. Chemical Research in Toxicology. 2010;**23**(4):733-739. DOI: 10.1021/tx900203v

[46] Soni N, Prakash S. Efficacy of fungus mediated silver and gold nanoparticles against *Aedes aegypti* larvae. Parasitology Research. 2012;**110**(1):175-184. DOI: 10.1007/s00436-011-2467-4

[47] Lamsal K, Kim SW, Jung JH, Kim YS, Kim KS, Lee YS. Inhibition effects of silver nanoparticles against powdery mildews on cucumber and pumpkin. Mycobiology. 2011;**39**(1):26-32. DOI: 10.4489/MYCO.2011.39.1.026

[48] Pineda L, Chwalibog A, Sawosz E, Lauridsen C, Engberg R, Elnif J, et al. Effect of silver nanoparticles on growth performance, metabolism and microbial profile of broiler chickens. Archives of Animal Nutrition. 2012;**66**(5):416-429. DOI: 10.1080/1745039x.2012.710081

[49] Ngô C, Van, De Voorde M. Nanotechnology in a Nutshell: From Simple to Complex Systems. Amsterdam: Atlantis Press; 2014. 1-495 p. DOI: 10.2991/978-94-6239-012-6-1

[50] Nanotea [Internet]. 2007. Available from: http://www.nanotechproject.org/cpi/products/nanotea/ [Accessed: 2018-05-30]

[51] Nanoceuticals™ Slim Shake Chocolate [Internet]. 2007. Available from: http://www.nanotechproject.org/cpi/products/nanoceuticalstm-slim-shake-chocolate/ [Accessed: 2018-05-30]

[52] Canola Active Oil [Internet]. 2007. Available from: http://www.nanotechproject.org/cpi/products/canola-active-oil/ [Accessed: 2018-05-30]

[53] HydraCel [Internet]. Available from: https://behealthyeveryday.eu/en/hydracel/ [Accessed: 2018-30-05]

[54] Nano Calcium [Internet]. 2010. Available from: http://natu-health.com/healthy-bones--and-joint/nano-calcium.html [Accessed: 2018-05-30]

[55] US patent—Production of carotenoid preparations in the form of coldwater-dispersible powders, and the use of the novel carotenoid preparations [Internet]. 1996. Available from: https://patents.google.com/patent/US5968251 [Accessed: 2018-05-30]

[56] LycoVit® [Internet]. Available from: https://nutrition.basf.com/en/Human-nutrition/LycoVit.html [Accessed: 2018-05-30]

[57] NovaSOL Curcumin [Internet]. 2014. Available from: http://novasolcurcumin.com/ [Accessed: 2018-05-30]

[58] Company Overview of NutraLease Ltd. [Internet]. 2018. Available from: www.bloomberg.com/research/stocks/private/snapshot.asp?privcapId=7222521 [Accessed: 2018-05-30]

[59] EUR-lex. Access to European Union Law [Internet]. 2018. Available from: https://eur-lex.europa.eu/homepage.html [Accessed: 2018-05-10]

[60] Amenta V, Aschberger K, Arena M, Bouwmeester H, Moniz FB, Brandhoff P, et al. Regulatory aspects of nanotechnology in the agri/feed/food sector in EU and non-EU countries. Regulatory Toxicology and Pharmacology. 2015;73(1):463-476. DOI: 10.1016/j.yrtph.2015.06.016

Chapter 5

Stabilization of Food Colloids: The Role of Electrostatic and Steric Forces

Camillo La Mesa and Gianfranco Risuleo

Additional information is available at the end of the chapter

http://dx.doi.org/10.5772/intechopen.80043

Abstract

The role that some forces exert on food colloid stability is discussed. The focus is on the combination of different energy terms, determining particle-particle attraction or repulsion. The forces are relevant in dispersion stabilization and macroscopic phase separation. The observed features depend on the energies at work and colloid concentration. Examples deal with food manipulations giving cheese, yogurt, and mayonnaise. All products result from the overlapping of forces jointly leading to aggregation or phase separation in foods. The combination of attractive, van der Waals (vdW), and repulsive, double-layer (DL) forces results in the dominance of aggregation or dispersion modes, depending on the particle concentration, on the force amplitude, and on their decay length. DL and vdW forces are at the basis of Derjaguin-Landau-Verwey-Overbeek (DLVO) theory on colloid stability. That approach is modified when these forces, jointly operating in bio-based colloids, overlap with steric stabilization and depletion modes. Steric effects can be strongly dispersive even at high ionic strength, despite this is rather counterintuitive, when depletion ones favor the nucleation in a single phase.

Keywords: van der Waals forces, double-layer forces, Poisson-Boltzmann equation, steric stabilization, depletion, food-based colloids

1. Introduction

Optimization of food properties is of fundamental interest because of growing demand for its widespread availability [1]. Substantial efforts tend to optimize the steps required in advanced food chain. Scientists and technicians focus on all preparation stages, from the collection of raw matter to transformation in the required form. Freezing, cooking, drying, salting, and all procedures which are part of human knowledge since thousand years are considered [2–6].

IntechOpen

Protein-rich preparations, such as anchovy paste [7], stockfish, dried venison, and cheese [8], are relevant examples. There is an urgent need to render old-fashioned preparations reliable and safe to a huge number of potential users. It is also necessary to ensure them good quality, together with homogeneous and "permanent" textures. Old-dated preparations give high-quality matter but operate in small scale, with drawbacks due to costs and durability. These preparations must be optimized to fulfill industry and safety requirements. Foods shall be stable for long times, still retaining their peculiar quality and taste. This is one of the reasons why modern preparations use stabilizers [9]. Storage must not require conditions hardly at hand in developing countries; think of the lack of low temperature and storage chains.

The focus is on semifluid matrices, such as creams and pastes, in other words, on items arbitrarily defined as *soft matter food*. Such products (ice creams, mayonnaise, pastes, sauces, etc.) are stabilized by addition of salts, lipids, proteins, and/or polysaccharides [10, 11]. Stabilizers are taste-neutral, fully biocompatible, and not expensive. They are obtained in large amounts, from the same sources as the products to be stabilized. In addition, food colloids must have peculiar rheological properties.

We do not consider explicitly the biological quality of preparations, which must fulfill the standards required from national/international panels. Many chemical, biochemical, and physicochemical properties characterize the features of stabilizers used in the food industry. Such properties always imply the stabilizer's capability to adsorb onto surfaces [12–14]. The latter is

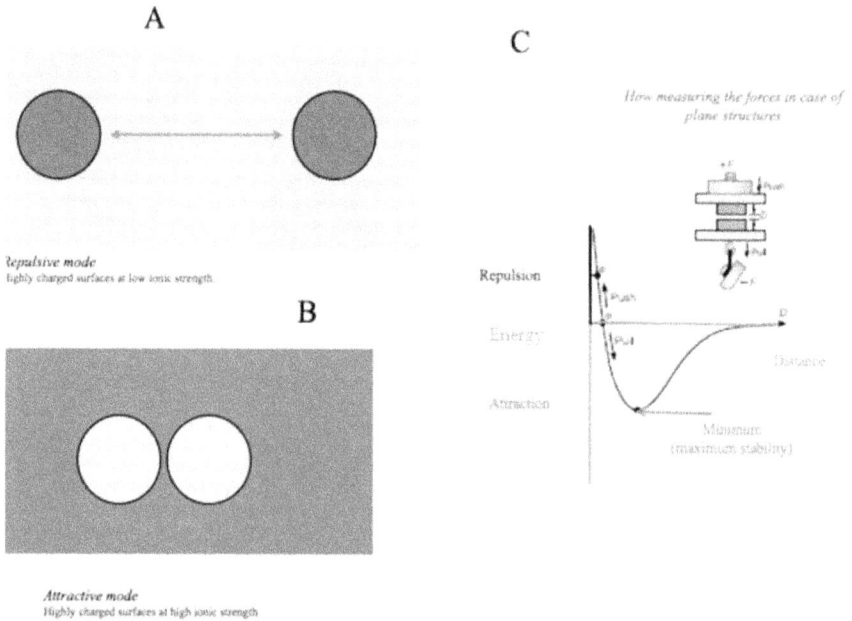

Figure 1. Electrolyte-modulated interactions among charged colloids. In (**A**), the case where repulsive electrostatic effects dominate is drawn; (**B**) refers to the reverse case (at high ionic strength). Light or dark blue colors indicate low and high ionic strength media, respectively. (**C**) The experimental determination of attractive/repulsive forces as a function of distance among two surfaces. The green horizontal line in that graph indicates equilibrium, i.e., $\Delta G = 0$.

the result of wrapping [15], steric [16, 17], osmotic [18], and electrostatic effects [19] and combinations thereof [20]. It is hard to ascertain whether the desired effect is due to the combination of more contributions. Think of the role played by proteins and polysaccharides as food stabilizers!

Electrostatic effects due to such stabilizers are relevant in most cases considered here. With this in mind, we report on the role that some forces exert in food stabilization. To proceed along this line, it is required to know the fundamental aspects of food biocolloids, interpreted according to the so-called Derjaguin-Landau-Verwey-Overbeek (DLVO) theory [21, 22]. In its original form, it is simple to handle and applies to all colloid mixtures, irrespective of their nature and physical state. DLVO theory combines attractive, van der Waals (vdW), and repulsive, double-layer (DL) forces. Refinements and modifications of the original theory are available [23–27]. The theory explains why food colloids remain dispersed, or coagulated, depending on the experimental conditions. This is because DL counteracts with vdW terms and their combination tunes the interaction modes. Similarly charged surfaces undergo long-range repulsions, and the energy barriers keeping them apart may be several K_BT units high [28]. However, if the electrolyte concentration in the medium increases, a secondary minimum in force *vs.* distance plots is observed (see **Figure 1**) [29]. The repulsive forces are minimized, and attractive ones dominate, i.e., coagulation occurs as σ approaches zero.

2. Some food preparation procedures

We report first on qualitative descriptions of food-making procedures. For some of them, the role that physical forces exert in the stabilization or phase separation is evident, in others much less. Enzymatic reaction steps are common. Energy barriers must be overcome to make the required processes effective; that is the reason for the need of heating during some preparation steps. Food preparation may occur in one or more stages.

Though they share the same raw product in common, significant difference occurs between cheese-making and yogurt-making procedures, briefly reported below. The former proceed by controlling milk fermentation, to get a product with specific organoleptic requirements in terms of appearance, flavor, taste, and texture. Such properties must be reproducible every time cheese is made. In fact, a particular cheese needs a specific preparation. In modern industrial cheese-making, the craft elements are retained to some extent, but there is more science than craft. In contrast, individual cheese-makers and craft-based factories operate on small scale and sell "handmade" products. In cases of the like, each batch may differ from another, as commonly occurs in the manipulation of natural products.

Some cheeses are deliberately left to ferment under the action of spores and bacteria; this leads to products of high added value in a niche market, such as *Roquefort*. In culturing the cheese-maker brings pasteurized milk in the vat to the thermal range promoting the growth of bacteria that feed on lactose. That sugar ferments into lactic acid. Bacteria may be wild, with non-pasteurized milk, added from a given culture, frozen or freeze-dried concentrates. Those producing only lactic acid are homofermentative; the ones producing CO_2, alcohol, aldehydes, ketones, etc. are heterofermentative.

Both homo- and heterofermentation produce cheeses with typical features in terms of taste, macroscopic textures, consistency, elasticity, presence of bubbles, and bubble size. When the cheese technicians judge that enough lactic acid has been developed, they add *rennet*, which precipitates casein. *Rennet* contains *chymosin* which converts κ-*casein* to *para-κ-caseinate*, the main component of cheese curd (see **Figure 2**). There is also a *glycomacropeptide*, almost always lost in the cheese *whey*. After adding the *rennet*, milk is left to form curds over a period of time. As curds are formed, milk fat is trapped in a casein matrix; *whey* must be released once cheese curds are fully developed. There are several ways to do that.

The presence of water and bacteria encourages further decomposition. Therefore, water or *whey* must be removed. When cheese curds are formed, a partial, sometimes significant, dehydration occurs. This gives rise to good quality products keeping their main features over time. In a stage termed *cheddaring* (from cheddar), curd acidity increases. When it has reached the required level, the curd is milled in pieces, and salt is added to arrest acid development. After some other stages, pressed cheese blocks are removed from the *molds* and waxed or stored for maturation. Vacuum packing removes O_2 and prevents fungal growth during maturation. This process is desired or not, depending on the required product. By going through a series of maturation steps where temperature and humidity are controlled, the cheese-maker allows the surface *mold* to grow and mold ripening of cheese by fungi to occur. Mold-ripened cheeses mature quickly compared to hard ones (weeks vs. months or years), because fungi are

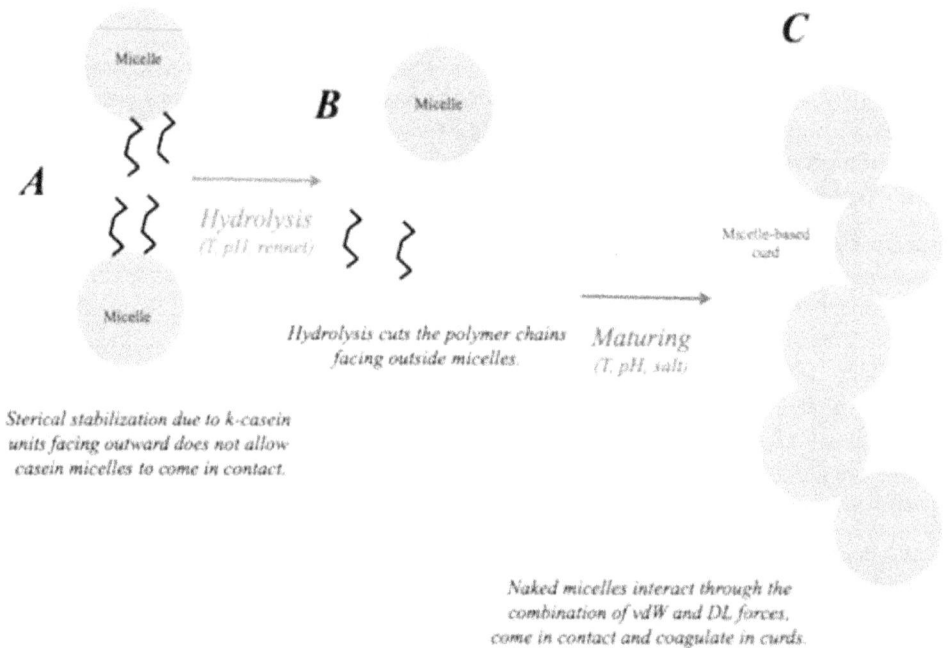

Figure 2. (**A**) Steric stabilization of casein micelles due to proteins facing toward the bulk and keeping micelles apart. (**B**) Enzymes present in the *rennet* cut κ-*casein* portions facing toward the bulk. (**C**) Depleted micelles attract each other and coagulate in curds. The process ends in cheese formation.

more active than bacteria. *Camembert* and *Brie* are surface-ripened by molds; *Stilton* is ripened internally and admits air to promote mold spore germination and growth. Surface ripening of some cheeses may be influenced by yeasts, contributing to flavor and coat texture. Others develop bacterial surface growths, giving characteristic colors and appearances.

Yogurt, conversely, is produced by bacterial fermentation of milk. Lactic acid acts on milk proteins and imparts yogurt its texture and flavor. Cow's milk is the common source to make yogurt; it may be homogenized or not. Yogurt is produced by *Lactobacillus delbrueckii* subsp. *bifidobacteria* (LDsB), *Lactobacillus bulgaricus*, and *Streptococcus thermophilus* bacterial cultures. Genome analysis of LDsB indicates that the bacterium presumably originated on the surface of a plant. Milk may have been exposed to contact with such plants, or bacteria transferred from domestic milk-producing animals. The real origin of yogurt preparation procedures is unknown but reasonably dates back to 5000 BC. To produce along, milk is heated to denature milk proteins so that they do not form curds. After cooling, the bacterial culture is mixed in, and the temperature is maintained for some hours to allow fermentation.

Mayonnaise, conversely, is a very peculiar product in terms of origin, components, and physical state. On physicochemical grounds, it is a surface-stabilized oil/water (o/w) dispersion, whose quality is determined by the presence of adsorbed lecithins at the o/w interface. The stability of this dispersion is modulated by tiny amounts of acetic or citric acid, which impart phospholipids as a moderate and permanent charge. Stability of the dispersion is modulated by added electrolyte, such as NaCl. In addition, the o/w dispersion, as a whole, adsorbs significant volume fraction of air. Thus, a heterogeneous two-phase dispersion acts as air dispersant; the final result is a three-phase system stabilized by surface-adsorbed lipids.

3. Some aspects of food colloids

3.1. General considerations

Animal-based foods and most of our own body organs generally contain about 55–75% water; in vegetables it can be over 90 wt%. Solid moieties are proteins, fats, lipids, etc., associating in different forms to give gel-like, liquid crystalline, amorphous, or semisolid matrices. Most tissues result from colloid packing. From that evidence comes the generalization that animals and vegetables are made of several different colloid entities, nicely, but functionally, interconnected. In all these systems, disperse colloid particles coagulate. Coagulation does not occur when particles are similarly charged; that is, coalescence is prevented by electrostatic forces. This holds also in dispersions of oil droplets stabilized by a phospholipid layer. Low amounts of electrolyte ensure lipid-covered droplets to repel each other. If the surface charge density, σ, or the related potential, Ψ, is moderate, the energy barrier among particles, proportional to $ze\Psi$, is low, and there is a marked tendency to coagulation. The limit at which such phenomena occur is known as coagulation/flocculation threshold.

Further increase of salt reduces Ψ and ensures permanent coagulation. As σ approaches 0, the DL force is null, electrostatic effects vanish, and the whole energy coincides with the vdW one; thus, particle-particle interactions become attractive. Similar conditions are met when food colloids aggregate in early manipulation stages and then redisperse as the pH or the ionic strength (I) varies. A simple case deals with oil droplets. The case of raw milk manipulation is substantially different from what is described above and ends in cheese formation. The whole process is controlled by the presence of fatty acids and glycerides existing as droplets; micelle-forming casein; coagulating enzymes, salts, and lactose (a milk sugar transformed in lactic acid) [30]; and so forth. The whole process is completed when aggregation/gelation occurs [31] and is governed by heating, enzymic activity, changes in pH, presence of ions, and combinations thereof.

Although casein micelles are charged, significant amounts of added salt do not ensure coagulation to form cheese seeds [32]. In fact, casein micelles are stabilized by steric effects, not allowing them to come in contact and coagulate. Steric stabilization counteracts attractive vdW forces and does not allow seed clustering. Such effects are minimized by the action of enzymes, cutting the κ-casein parts facing outward micelles. In the early cheese-making steps, pH activates/deactivates hydrolytic enzymes [33], whose activity also depends on T [34].

The presence of *rennet*, essential in the first stages of cheese curdling, is also relevant. Ion content and valence (calcium better suits compared to monovalent ions) favor casein aggregation in large micelles and, therefore, curd formation [35, 36]. To elucidate such aspects, we introduce below an approach to electrostatic stabilization and show that it, in combination with vdW forces, is relevant in food formulations, as indicated in **Figure 3**.

3.2. Electrostatic forces

Colloid entities are characterized by a given mass density and average size, can be more or less size polydisperse, and wear a surface electrostatic potential [37]. When dispersed in water, uncharged colloids readily coagulate, but surface charge density avoids that process. Irrespective of their nature and shape, colloid particles are covered with stabilizers, adsorbing thereon, and imparting them a permanent surface charge. In consequence of that, particles repel, depending on the modulus of Ψ, |Ψ|, which exerts a long distance effect and scales with kD (**Figure 3**).

The distance is D and 1/k is Debye's screening length. Repulsion occurs when particles are close to each other. The effect has the same meaning as that between planar surfaces of equal Ψ values (**Figure 4**).

Electrostatic potentials decay according to

$$\Psi(x) = \Psi^{\circ} \exp^{-kD} \tag{1}$$

where D is the distance from a charged surface of nominal potential equal to Ψ°. The meaning of k has been given above.

Another master equation for electrostatics refers to the interaction between two surfaces characterized by the same Ψ. It decays according to

$$\nabla^2\Psi = d^2\Psi/dx^2 + d^2\Psi/dy^2 + d^2\Psi/dz^2 = -\left(\varrho/\varepsilon\varepsilon^{\circ}\right) \tag{2}$$

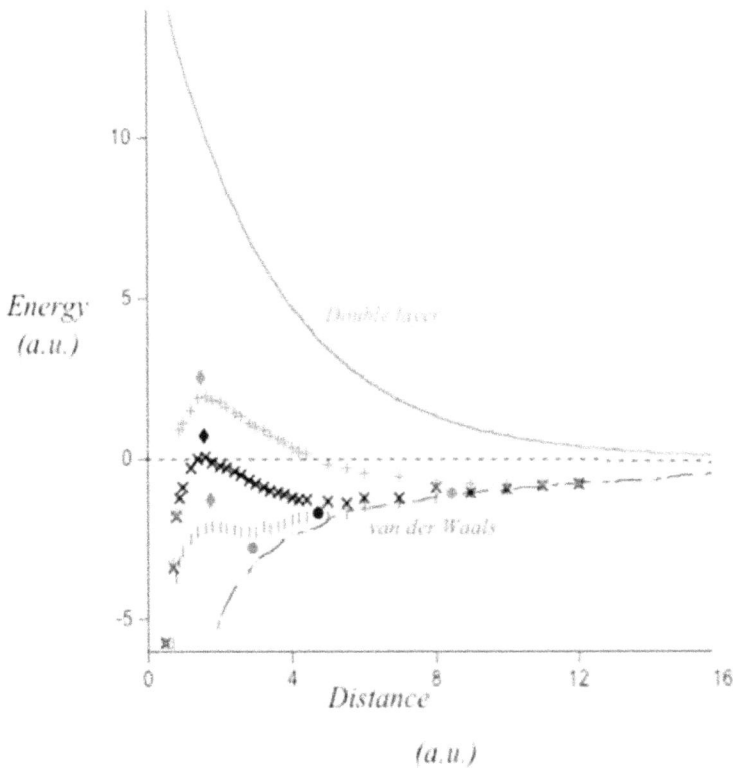

Figure 3. Combination of vdW and double-layer (DL) forces as a function of distance among two particles for high (red), medium (black), and low (l) surface charge densities. vdW terms are always attractive, i.E., E < 0; DL ones are always repulsive. Their combination results in energy vs. distance plots. The location of maxima, ◆, and minima, ●, depends on ionic strength, I. The primary minima at very short distances are not indicated. The maxima in the curves represent the location of energy barrier.

where ϱ is the ion number density of the medium and ε and ε° are the permittivity of vacuum and of the dispersant, respectively. The electric field is radial and its value does not depend on the direction; thus, we consider its components along only one axis, say x. And, Eq. (2) can be rewritten as

$$\nabla^2\Psi = d^2\Psi/dx^2 = -(\varrho/\varepsilon\varepsilon^\circ) \tag{3}$$

Let us consider now the statistical energy terms. Boltzmann's law for the distribution of charged species in a given medium can be written as

$$c_i = c_i^\circ \exp^{-(ze\Psi/KBT)} \tag{4}$$

where c_i is the local concentration of the ith ion, c_i° is its equilibrium value, $ze\Psi$ is the energy associated to the electric field for an ion of valence (z), and K_BT is the thermal one. Eq. (3) is

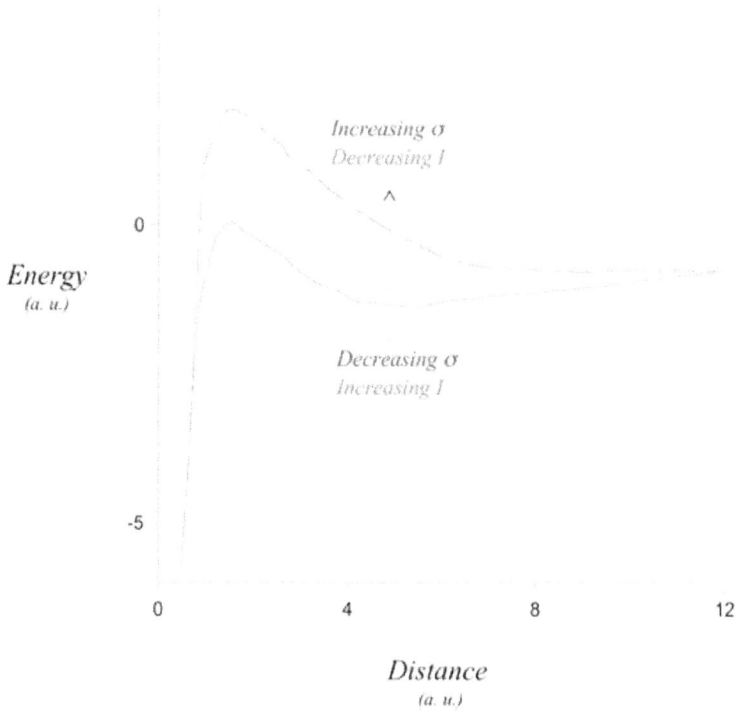

Figure 4. Relationship between ionic strength, surface charge density, and attractive/repulsive forces at fixed D. It is evident that repulsive forces turn to attractive, depending on I value.

modulated by the electrical to thermal energy ratio. The balance of such forces determines the spatial distribution of ions around a charged entity, depending on the electric field and thermal motions. Then,

$$\varrho = \varepsilon(c^+ - c^-) = \varepsilon c^\circ \left[\exp^{-(ze\Psi/KBT)} - \exp^{(ze\Psi/KBT)} \right] \tag{5}$$

In the above form, the equation (usually different from 0) represents the local charge density due to an ion in excess. If $|ze\Psi/K_BT|$ is $\ll 1$, the difference between exponents can be transformed in hyperbolic form ($\exp^x - \exp^{-x} = 2\sinh x$) and linearized. We assume $x = ze\Psi/K_BT$. Thus, when $x \ll 1$, Eq. (4) indicates a *linear* perturbation regime. Such conditions are currently used to determine the electrostatic energy contributions. Advantages due to linearization are substantial.

The charge density, ϱ, is related to the surface potential, σ, which, in turn, depends on Ψ. The links between ϱ, σ, and Ψ are expressed as

$$\sigma = -\int \varrho dx \tag{6}$$

$$\sigma = (2n^\circ \varepsilon K_B T/\Pi)^{1/2} \sinh(ze\Psi/K_B T) \tag{7}$$

where ε is the dielectric permittivity of the medium. σ relates the system energy to electrical terms, according to

$$\Delta G = -\int \sigma d\Psi \qquad (8)$$

Let us consider the role of electrostatic forces, favoring/disfavoring phase separation. In simple cases the focus is on the formation of mayonnaise and yogurt; subsequently, the more cumbersome case of cheese is described. vdW forces are present in all such cases. Significant differences arise when steric, osmotic, and DL contributions counteract vdW forces, do not allow adhesion, or shift the coagulation threshold to high concentrations. These features, observed in some cheese-making stages, are outlined below.

3.3. Electrostatic vs. vdW forces

In the classical formulation of DLVO theory, vdW forces are combined with DL ones. For bodies at constant T, the interaction energy (E_{int}) significantly depends on distance (D). At high D values, E_{int} is zero and all contributions vanish. Modulation of the above terms results in the presence of a primary and a secondary minimum. The first one occurs at very short distances, and the second one, at higher ones. The secondary minimum shifts to lower values in proportion to I; an energy barrier separates it by the primary minimum. The barrier height is related to the activation energy of coagulation (**Figure 3**).

The secondary minimum in **Figure 3**, some $K_B T$ units high, shifts to lower distances in proportion to I. The tendency to coagulate is represented by the progressive overlapping of vdW and DL curves. The minimum at short distances is not indicated; the maximum is related to E_{att}. The role of ionic strength can be evidenced considering the electrostatic potential among two surfaces with a fixed number of charges per unit area which are shielded by increasing concentrations of salt (**Figure 4**).

In **Figure 5** we indicate how the electrostatic potential changes with I. In distilled water, $\Psi°$ rapidly increases with ion concentration. Neutral electrolytes (in the concentration range of 10^{-3} moles kg^{-1}) have a buffer effect on $\Psi°$. Since most foods contain substantial amounts of salt, the region where the effect of Ψ is significant ranges from 25 to 100 mV, in modulus [38]. For values <|25| mV, samples tend to coagulate; above 100 mV counterion adsorption becomes large, thus minimizing electrostatic repulsions. Small changes in Ψ values are large enough to ensure dispersion, aggregation, or sedimentation in all such media. That is why a careful balance of surface energy terms is necessary.

The electrostatic contributions in real systems are measured by the so-called ζ-potential, a distance (d) apart from the colloid particle surface, at the slipping plane limit. ζ-potential is measured by electrophoretic mobility experiments or laser Doppler methods [39]. The decay of ζ-potential with pH and/or I is easily determined (see **Figure 6**).

Accordingly, surface charges are titrated, and coagulation or redispersion occurs (**Figures 5 and 6**). The salient point in the latter is the zero surface charge value, where precipitation

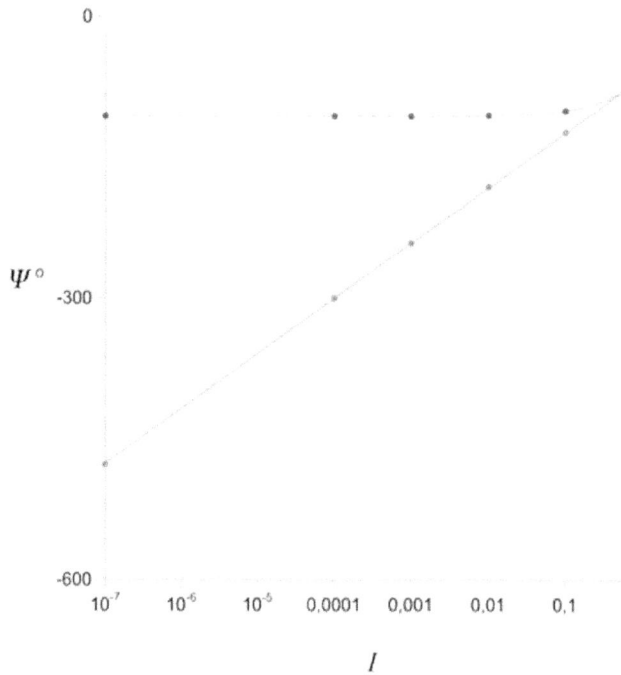

Figure 5. Effect of ionic strength, I, in moles kg^{-1}, on the electrostatic potential, $\Psi°$, of particles dispersed in water, red color line, and in 3.0×10^{-3} moles kg^{-1} MgCl$_2$, at 25.0°C. Note that I in bio-based systems can be significant; usually, the overall salt concentration is $>5.0 \times 10^{-3}$ moles kg^{-1}.

occurs. The electrostatic theory, thus, explains why salts screen repulsions, in direct proportion to valence and concentration. For instance, oil droplets covered by a charged lipid layer coagulate when NaCl content in the dispersing water-based medium reaches a critical value (some millimoles kg^{-1}).

Particle size, disparity, shape, and physical state (i.e., solid- or liquid-like) are immaterial. Although its value seems moderate, the surface charge density is relevant. σ of lipid-coated oil droplets (**Figure 5**) is about 1 unit charge/15 nm^2. And, despite such a relatively low value, stabilization is effective. Also, ion valence is relevant, as indicated by the relation $I = 1/2 \ \Sigma_{i\ =\ 1}$ $c_i z_i{}^2$. The coagulation concentration is concomitant to the secondary minimum of the curves in **Figure 3** and depends on z_i. The combination of such effects is also responsible for the stabilization effect due to proteins. Solving the above questions and taking into account protein stabilization are relevant in steric stabilization, as outlined in the forthcoming section.

3.4. Steric stabilization

This concept applies to entities covered by polymers or polyelectrolytes protruding outside the surfaces on which they are bound. Coatings consist of covalently linked (CL) or physically adsorbed (PA) polymers: the differences among two such classes are energetically significant.

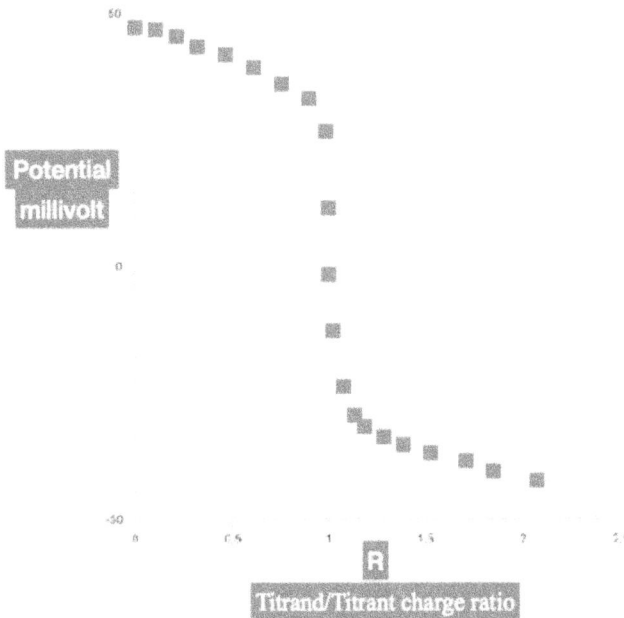

Figure 6. Dependence of ζ-potential, in mV, on the ratio titrant to titrand, R. When R ≅ 1, the ζ-potential approaches zero. The curve is symmetrical with respect to R. Data refer to 3.52×10^{-2} dispersions of sunflower oil (in volume fraction) with 2.3 mg/ml dipalmitoylphosphatidylcholine (DPDC) as dispersant. The solvent is 2.50×10^{-2} moles kg^{-1} NaCl and the temperature is 35.0°C. In this case the titrand/titrant ratio, R, depends on pH.

CL entities have permanent stabilizer/particle ratio. PA ones partition between particle surface and bulk, depending on the system composition and affinity; these composites are nonstoichiometric. PA surface adsorption energies are grossly one order of magnitude lower than those pertinent to covalent ones; indeed, both ensure substantial stability. The mentioned modes and energies have both advantages and drawbacks. CL polymers are depleted from the particles' surface by chemical reaction. This holds, for instance, when κ-*casein* is cut away from the surface of micelles by the combined action of *rennet, pH*, and temperature. Otherwise, casein micelles are stable for an indefinitely long time. The hydrolytic capacity of *rennet*, thus, is a prerequisite for effective coagulation. Once hydrolysis has gone to completion, PAs are detached from casein micelles and partition with the bulk. Unbalanced osmotic effects due to bulk polymer concentration in excess result in *depletion flocculation* [40]. That is, PAs are released and no longer stabilized micelle clusters of casein and calcium phosphate coagulate, and phase separation occurs to give cheese seeds and clusters.

Steric stabilization overlaps with other effects, jointly tending to keep particles dispersed. The result is modulated by the presence of charges on the protruding polymers. Osmotic, electrostatic, steric, and hydration forces sum each other and counteract vdW ones (**Figure 7**).

It is clear, thus, why particles may remain dispersed even when DL contributions are minimized. In some instances, the terms due to the aforementioned forces may be noticeable and

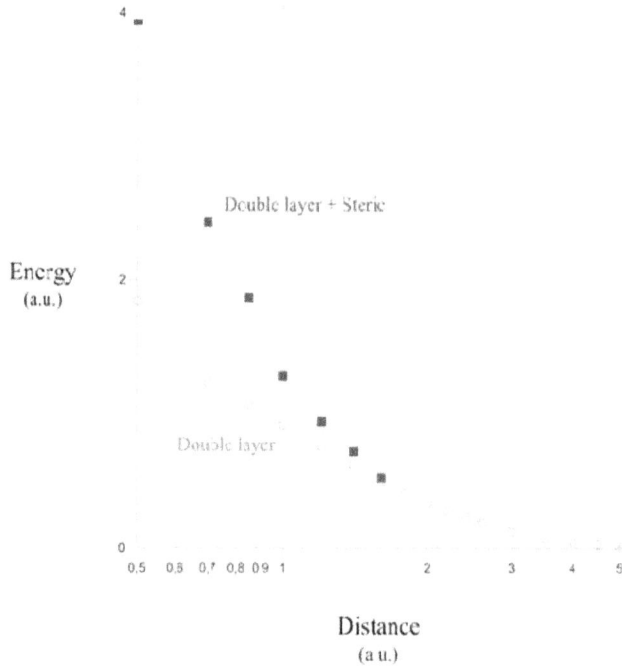

Figure 7. Dependence of the interaction energy, E, in arbitrary units, on the normalized distance among particles, D, in case of double layer, red, or when double layer and steric contributions overlap.

favor the dispersed state with respect to coagulation. For instance, consider the possibility to add to the plot in **Figure 7** osmotic and hydration forces.

We may combine all forces effective in a given medium in the generalized relation:

$$E_{tot} = \sum_{i=1} E_i \exp^{-k_iD} \tag{9}$$

where E_i is a given energy mode, D is the distance, and $1/k_i$ is the related screening length. Expectedly, forces decay exponentially, even though this statement is not to be generalized. Most forces, in fact, scale as $1/D^n$ (with $n \geq 3$). Eq. (8) conforms to short-distance decay modes and indicates that repulsion rapidly decreases with distance [20]. Attraction, conversely, is governed by vdW terms, responsible for phase separation. In this regard, the differences between yogurt and cheese coagulation clarify which forces govern the onset of such materials.

4. Conclusions

Forces responsible for attractive/repulsive interactions among food colloids are discussed. One must be aware that some of them, i.e., DL ones, are ubiquitous, although not always stabilizing. This ineludible fact is due to the presence of ions in most media. That is why association or phase separation is common as the ionic strength increases. Salting/desalting methods are responsible for cheese- and yogurt-making, among others. It is worth noting that the latter occurs in media which *per se* contain ions in the original matrix. It is also worth mentioning that calcium is present in significant amounts; that is why aggregation is relatively easy. Additional effects counteracting vdW ones arise from steric and osmotic contributions. It was suggested how to face with such processes and how to estimate from simple considerations stabilizing/destabilizing effects. It must be pointed out that DL effects significantly reduce on increasing the ionic strength.

The hierarchy of active forces and their combination ensure slightly different aggregation modes, giving more or less complex conglomerates. These can be homogeneous or not, depending on the nature of dispersed colloids. The supramolecular phases thus obtained may be due to intertwingled association of one colloid type into matrices made up by another. The uptake of fat droplets in cheese curds is a pertinent example. More effects, mostly due to surface adsorption, may significantly affect the final quality and appearance of the mentioned *soft matter* food.

Perspectives are governed by the continuous developments and optimization of food manipulation processes. These usually optimize former know-how, not disregarding the maintenance and quality of classical produce. In this regard it is safer to rely on craft-tempered goods produced by advanced technologies.

Author details

Camillo La Mesa[1]* and Gianfranco Risuleo[2]

*Address all correspondence to: camillo.lamesa@uniroma1.it

1 Department of Chemistry, Cannizzaro Building, La Sapienza University in Rome, Italy

2 Charles Darwin Department of Biology and Biotechnology, La Sapienza University of Rome, Italy

References

[1] Martínez-Guido SI, Betzabe Gonzales-Campos J, El Halwagi MM, Ponce-Ortega JM. Sustainable optimization of food networks in disenfranchised communities. ACS Sustainable Chemistry & Engineering. 2017;**5**:8895-8907

[2] Marotz LR. Health, Safety, and Nutrition for the Young Child. Wadsworth Publishing Co.; 2008. ISBN: 978-1-4283-2070-3

[3] Matthews KR. Review of Published Literature and Unpublished Research on Factors Influencing Beef Quality. EBLEX R&D UK Agric. Horticulture Develop. Board; 2011

[4] Richardson M, Matthews K, Lloyd C, Brian K. Meat Quality and Shelf Life. Better Returns Progr. EBLEX Agric. Horticulture Develop. Board; 2012

[5] U.S. Dept. Agric., Food Safety Inspec. Serv. Fact Sheet: Freezing and Food Safety. 2010

[6] University of Missouri. Vegetable Harvest and Storage. 2011

[7] Dalby A. Food in the Ancient World from A to Z. Taylor & Francis; 2013. ISBN: 978-1-135-95422-2

[8] Asher D. The Art of Natural Cheesemaking. Vermont: Chelsea Green Publishing; 2015

[9] Imeson A. Food Stabilisers, Thickeners and Gelling Agents. Wiley; 2011. ISBN: 978-1-4443-6033-2

[10] Freybler LA, Gray JI, Asghar A, Booren AM, Pearson AM, Buckley DJ. Nitrite stabilization of lipids in cured pork. Meat Science. 1993;**33**:85-96

[11] Pittia P, Paparella A. Safety by control of water activity: Drying, smoking, and salt or sugar addition (Chapter II). In: Prakash V, Martin-Belloso O, Keener L, Astley SB, Braun S, McMahon H, Lelieveld H, editors. Regulating Safety of Traditional and Ethnic Foods. 2016. pp. 7-28. ISBN: 978-0-12-800605-4

[12] Benichou A, Aserin A, Garti N. Protein-polysaccharide interactions for stabilization of food emulsions. Journal of Dispersion Science and Technology. 2002;**23**:93-123

[13] Nambam JS, Philip J. Competitive adsorption of polymer and surfactant at a liquid droplet interface and its effect on flocculation of emulsion. Journal of Colloid and Interface Science. 2012;**366**:88-95

[14] Zhu G, Mallery SR, Schwendeman SP. Stabilization of proteins encapsulated in injectable poly (lactide-co-glycolide). Nature Biotechnology. 2000;**18**:52-57

[15] Schott H. Saturation adsorption at interfaces of surfactant solutions. Journal of Pharmaceutical Sciences. 1980;**69**:852-854

[16] Wilson CJ, Clegg RE, Leavesley DI, Pearcy MJ. Mediation of biomaterial-cell interactions by adsorbed proteins: A review. Tissue Engineering. 2005;**11**:1-18

[17] Zeisig R, Shimada K, Hirota S, Arndt D. Effect of steric stabilization on macrophage uptake in vitro and on thickness of the fixed aqueous layer of liposomes made from alkylphosphocholines. Biochimica et Biophysica Acta - Biomembranes. 1996;**1285**:237-245

[18] Webster AJ, Cates ME. Osmotic stabilization of concentrated emulsions and foams. Langmuir. 2001;**17**:595-608

[19] Perutz MF. Electrostatic effects in proteins. Science. 1978;**201**:1187-1191

[20] Israelachvili JN. Intermolecular and Surface Forces. 3rd ed. Academic Press; 2015

[21] Derjaguin B, Landau L. Theory of the stability of strongly charged lyophobic sols and of the adhesion of strongly charged particles in solutions of electrolyte. Acta Physicochimica U.R.S.S. 1941;**14**:633-640

[22] Verwey EJW, Overbeek JTG. Theory of the Stability of Lyophobic Colloids. Elsevier; 1948

[23] Boström D, Franks GW, Ninham BW. Extended DLVO theory: Electrostatic and non-electrostatic forces in oxide suspensions. Advances in Colloid and Interface Science. 2006; **123–126**:5-15

[24] Grasso D, Subramaniam K, Butkus M, Strevett K, Bergendahl J. A review of non-DLVO interactions in environmental colloidal systems. Reviews in Environmental Science and Biotechnology. 2011;**1**:17-38

[25] Guldbrand L. Electrical double layer forces. A Monte Carlo study. The Journal of Chemical Physics. 1984;**80**:2221-2228

[26] Pashley RM. DLVO and hydration forces between mica surfaces in Li+, Na+, K+, and Cs+ electrolyte solutions: A correlation of double-layer and hydration forces with surface cation exchange properties. Journal of Colloid and Interface Science. 1981;**1981**(83):531-546

[27] Wennerstroem H. The cell model for polyelectrolyte systems. Exact statistical mechanical relations, Monte Carlo simulations, and the Poisson–Boltzmann approximation. The Journal of Chemical Physics. 1982;**76**:4665-4672

[28] Chen W. Energy barriers for thermal reversal of interacting single domain particles. Journal of Applied Physics. 1992;**71**:5579

[29] Pashley RM. Interparticulate forces (Chapter III). In: Laskowski JS, Ralston J, editors. Colloid Chemistry in Mineral Processing. Elsevier; 1992. pp. 97-114. ISBN: 0-444-88284-7

[30] Smit G, Smit BA, Engels WJM, Wim JM. Flavour formation by lactic acid bacteria and biochemical flavour profiling of cheese products. FEMS Microbiology Reviews. 2005;**29**: 591-610

[31] Singh A, Latham JM. Heat stability of milk: Aggregation and dissociation of protein at ultra-high temperatures. International Dairy Journal. 1993;**3**:225-237

[32] Hallèn E. Coagulation properties of milk. Association with milk protein composition and genetic polymorphism [Ph.D. thesis]. Uppsala: Faculty of Natural Resources and Agricultural Sciences; 2008

[33] Hotnida S, Nidhi B, Bhesh B. Effects of milk pH alteration on casein micelle size and gelation properties of milk. International Journal of Food Properties. 2017;**20**:179-197

[34] Dumpler J. Heat Stability of Concentrated Milk Systems. Springer; 2017. pp. 47-62. ISBN: 978-3-658-19696-7. Chapter III

[35] Ikonen T, Morri S, Tyriseva AM, Ruottinen O, Ojala M. Genetic and phenotypic correlations between milk coagulation properties, milk production traits, somatic cell count, casein content, and pH of milk. Journal of Dairy Science. 2004;**87**:458-467

[36] Tercinier L, Ye A, Anema SG, Singh A, Singh H. Interactions of casein micelles with calcium phosphate particles. Journal of Agricultural and Food Chemistry. 2014;**62**: 5983-5992

[37] Mills I et al. Definitions, terminology and symbols in colloid and surface chemistry. Pure and Applied Chemistry. 1972;**31**:579-638

[38] Adamson AW. Physical Chemistry of Surfaces. 5th ed. New York: Wiley; 1990. pp. 421-433. Chapter IX

[39] Tadros TF. Interaction forces between particles or droplets in agrochemical dispersions (suspension concentrates or emulsions, EWs) and their role in colloid stability. In: Colloids in Agrochemicals. Wiley; 2009. pp. 77-91. Chapter V

[40] Sing H. Milk proteins functionality in food colloids. In: Dickinson E, editor. Food Colloids: Interactions, Microstructure and Processing. RSC; 2005. pp. 179-193. ISBN : 0-85404-638-0

www.ingramcontent.com/pod-product-compliance
Lightning Source LLC
Chambersburg PA
CBHW081234190326
41458CB00016B/5778